APOLLO 11 SECRET

APOLLO 11 SECRET

NEW EVIDENCE ABOUT APOLLO 11

Francisco Villate

2024

Cover design: Rhal Zahi

English editor: Maureen Campanile

Copyright © 2024 Francisco Villate

Photos of Apollo Missions courtesy of NASA.

US Library of Congress Office registration: TXu 2-421-391

All rights reserved.

Published by Francisco Villate (rhalzahi.com)

ISBN: 978-1-0688417-1-2

CONTENTS

PROLOGUE

NASA did reach the Moon. But we were deceived. The story we know about the Apollo program is partially true. This book does not present another conspiracy theory within the vast stormy ocean of conspiracy theories regarding Apollo 11. Here, we offer facts and hypotheses supported by logic indicating at least two lies.

The intention of this book is not to challenge NASA's trips to the Moon. It does not seek to dispute Moon landings. On the contrary, it proves that NASA was successful, but... it also proves that we have been lied to. The facts indicate that Apollo 11 never landed on the Moon, and another Apollo mission, a mystery mission, later installed all the equipment that Apollo 11 should have installed.

As a researcher, I find myself between two extremes. One relates to those who support many conspiracy theories and say that NASA never went to the Moon. These conspiracy advocates rejected me when I tried contacting them because I told them the Apollo missions were successful. They sent me information to convince me that aliens are demons or that the Earth is flat and provided other unconventional theories. On the other side are the skeptical scientists, who rejected the idea that Apollo 11 never landed on the Moon. Upon hearing my findings based on facts, some left me and immediately labeled me a "Moon Hoaxer" because they did not want to look at the evidence. Similar to Galileo's skeptics, they "don't want to look through the telescope" to check the facts for themselves. Logic and facts, based on actual findings, are the basis of this book. At the same time, it presents these facts straightforwardly so that anyone can understand.

I have always been fascinated with the Moon. When I was a teenager, I often spent many hours watching it. I went out into the street to observe it, even through the windows of my house or on top of our home roof. I found many ways to contemplate the wonderful Moon in ecstasy. My brothers used to make fun of me and told me that I looked like the "Bull in love with the Moon," referring to the Spanish song about a bull that spends his nights gazing at the Moon, hopelessly enamored with it.

Also, since childhood, I have been very interested in space travel and astronomy. As a child and adolescent, I built small rockets using matches as fuel. I developed small two-stage rockets, some of them with parachutes. Once, at 13, I caught a fly and put it inside a tiny capsule of one of my rockets. It would be the pilot of my ship. The rocket would eject the capsule when it reached maximum altitude, and a small parachute I made from a plastic bag would open. At the moment of takeoff, my brother and I ignited the small rocket that carried the insect living inside the capsule. Unfortunately, the missile exploded at the launch pad. The rocket and capsule flew about 15 meters high and then fell. The parachute did not open completely, and the capsule and astronaut crashed into the ground. Concerned, we slowly opened the door of the tiny capsule to check the health of the fly. It was alive but dizzy. It came out of the capsule and, for a couple of minutes, did not move. Then it flew away in fright, perhaps cursing us for the unpleasant experience it had just gone through. From then on, I never again used any living beings in my experiments.

As the years went by and I reached adulthood, I joined a group of people interested in developing rockets, and together, we conducted several experiments. We designed and tested different types of gunpowder. I made a machine that used a pendulum with a small rocket held at one end, with additional weight as ballast. When I ignited the missile, the pendulum moved, and we learned its force depending on its amplitude. I developed the formulas to calculate that force. I planned to improve the mechanism of plotting the pendulum's elongation variation on a graph over time to determine the variation of trust. The mathematics and physics of that experiment fascinated me. I wanted to assess the changes in the rocket's thrust and weight to calculate the potential altitude it could reach. To do the experiments, a fellow group member fashioned a cylindrical bronze container in which we placed the powder for testing. I made a straightforward remote ignition mechanism with an electrical resistor that lit a match. This setup allowed us to safely ignite the small rocket from a distance.

We went on an excursion to a farm on the outskirts of Bogotá, Colombia, to make astronomical observations. We brought the apparatus and several gunpowder charges with us, intending to demonstrate their effects to the astronomy enthusiasts accompanying us in the middle of the night. However, the remote ignition system did not work correctly. Driven by the pressure of the moment and the spectators, I recklessly approached the machine and ignited the rocket myself. A couple of tests worked well. The rocket ignited, and I could measure the resulting elongation to calculate the thrust later. But on the third test, the missile did not ignite. I attempted to ignite it with a match but unintentionally ended up heating the brass part instead. And inside that metal cylinder were 4 grams of gunpowder, equivalent to a revolver bullet. The rocket exploded at an arm's length distance from my face. The bronze piece broke into four parts; one hit me in the left eyebrow. I felt as if someone had punched me in the head with a baseball bat. I took five steps back and fell, knocked out. Stunned, I saw everyone standing over me, frightened and worried about my health. I don't know what amazed me more, seeing them so scared or feeling very calm as if it wasn't tragic. I felt no fear, no pain, just tranquility.

We rushed to a doctor in the nearest town and woke him up in the middle of the night. He treated me kindly, stitching a four-centimeter-long wound on my eyebrow, and then noticed that a piece of metal might be lodged inside my eyelid, which was already inflamed, completely covering my entire eye. He removed the stitches and then, with a pair of tweezers, carefully tried to find it and pull it out. He could not catch it. He sewed up the stitches again, and we returned to the farm where the owners gave me a room to sleep in, with a comfortable bed. It was much better than the tent we had taken to camp. A great friend and fellow member of the astronautics group, who was doing our experiments, offered to share the room with me in case I required some support during the night. He was very concerned about my health and wanted to make sure nothing happened to me. I remember not being able to sleep that night, not because of any pain or health problems, but because my friend snored all night.

A week later, I went for an x-ray based on the doctor's recommendation. I wanted to confirm whether a small metal splinter was lodged in my eyelid. I still remember the screams of the X-ray machine operator as he developed the test's negatives. He was terrified by what he saw on the x-ray film. A large, metallic, twisted piece took up my entire eyelid. It had a sharp point that was only 2 millimeters from the surface of my eye.

My siblings, friends, and I rushed to a clinic. The optometrists who checked me there were amazed at what I had in my eyelid and that I had no ocular disturbance. They hospitalized me that night, and the next day, I went into the operating room, and they took out a 4-centimeter-long piece of bronze. Miraculously, I had no damage to my vision. The doctor who attended me told me that I was lucky because if that piece of metal had hit my eye or if it had hit my forehead, nothing would have stopped it from reaching my brain. It hit a thick skull area, which was denser and more resistant. After that incident, I abandoned my short career to reach space. I opted for a safer approach, dedicating myself to observing it from the ground through my telescope.

I continued with my studies of astronomy and astrophysics. Through studying the sky, learning the constellations, building telescopes, and observing the night sky daily, I had the opportunity to witness two UFO events. One was a luminous point, resembling a star, that danced before me and two other astronomy enthusiasts while we were engaged in astronomical observations. On another night, I watched some 11 to 13 bright objects flying, meeting in the sky, making abrupt turns, and then traveling in a "V" formation.

Because of my fondness for astronomy, people who had experiences with UFOs often sought me out for explanations about the phenomenon. I have had the opportunity to speak with and personally interview four contactees who claim to have been taken aboard an extraterrestrial craft and have talked to star travelers. All of them describe the extraterrestrials as humans similar to us. One of them, the most amazing, was Billy Meier. Based on his evidence, I conducted scientific research that indicated this case was genuine. I have written two books about it with a friend and associate, Christopher Lock, who teaches at a university in Osaka, Japan.

Why do I mention all this, and how does it relate to the Apollo program? Within what Billy Meier states, he includes an explanation[1] about the lies of the Apollo program explained to him by his extraterrestrial contacts. Ten years ago, I heard him in a report in the documentary "As the Time Fulfills,"[2] produced by Michael Horn. There, Billy mentioned that he was authorized to say that the Apollo 11 Mission did not land on the Moon and that perpetrators of a hoax deceived all humankind. He also stated that Apollo 13 was a hoax because it did not have the failure that forced them to abort the mission and that Apollo 13's mission was to deliver and set up the remaining equipment on the Moon that had not been deployed by Apollo 11 in the Sea of Tranquility. He said that for political reasons, to show the world and the Russians that the USA had advanced technology, they pretended that they had landed on the Moon during Apollo 11, but it was not so.

I had heard several theories about the Apollo 11 hoax before. Although the evidence presented to support the hoax theory did not convince me, it seemed likely to others. But to hear from Billy Meier that not only was Apollo 11 a hoax but that Apollo 13 was a ruse as well surprised me. I had not heard of such a theory before. Ten years ago, when researching the intended landing location of Apollo 13, I noticed it was located in a different region of the Moon. So, if Apollo 13 had landed at the Apollo 11 site during its mission, the Sun's position would have been too high, close to its zenith. They would not have been able to take pictures and videos to substitute the existing Apollo 11 ones, which, according to this theory, would have been faked. I have not investigated this event any further for several years.

My interest in the subject of Apollo 11 and 13 reawakened recently. My research revealed that Apollo missions could reach the Moon at varying speeds, either quickly or slowly. I noticed that Apollo 8 arrived in only two days and 20 hours, and other Apollo missions took more than three days. Then, I checked to see what would have happened if Apollo 13 had flown at the same speed as Apollo 8. I noticed it would have arrived at the Sea of Tranquility, Apollo 11's designated location, at a very similar lunar phase to what it was during Apollo 11. Consequently, the Sun would be approximately the same elevation above the lunar horizon. Had it arrived there, the photos and videos from Apollo 13 would be very similar to what we would expect from Apollo 11. I then found that Apollo 13, having more fuel, would have been able to reach the Moon very fast, even faster than Apollo 8.

The speed of a rocket flying to the Moon can vary. It depends on the amount of fuel used. In Earth orbit, the service module, attached to the Lunar Module and boosted by the third stage of the rockets, accelerates to escape Earth's gravitational field. If they fire the engine longer, the velocity will be higher. Then, it is the inertia that takes the spacecraft to the Moon. It is not required to give it more thrust. Still, it is necessary, when approaching the Moon, to turn around and restart the engines to slow down and enter its orbit. It must use additional fuel to reduce speed if it moves too fast.

So, it was possible that Apollo 13 could have traveled at a high speed and arrived "on time" at the Sea of Tranquility, landed there, and stealthily installed all the equipment that Apollo 11 had not installed before.

How do we prove that Apollo 11 never landed on the Moon? By interrogating the astronauts? In Billy Meier's contact notes about this subject, he says that the Apollo 11 astronauts, like the Apollo 13 astronauts, had false memories implanted into their minds through hypnosis techniques. In the case of Apollo 11, the hypnosis made them remember that they had landed on the Moon. In the case of Apollo 13, they recalled that they had failed and aborted the mission. It may sound like a science fiction movie. If it is true, then the astronauts will only talk about what they remember, even if their memories are false and forcibly implanted. It is tough to prove and will likely remain just another claim among the conspiracy theories surrounding the Apollo program.

However, by investigating the photographic and video evidence of Apollo 11, anyone can find several facts that can be scientifically proven to indicate that Apollo 11 did not land on the Moon. And if it did land there, Armstrong or Aldrin did not take the pictures we see of this mission. That's what this book is about. And it doesn't require "rocket science" to figure that out. A high school student can prove it.

This study will present verifiable facts and offer several hypotheses to explain them. Readers can develop their own theories in search of the truth.

I am not accusing NASA, as an institution, of lying to us about the Apollo missions. A small group of people connected with NASA carried out the hoaxes. NASA has been very successful in its space projects, and in particular, if the hypotheses raised here are true, the Apollo program had no failures. Its missions were successful despite being shrouded in controversies. It is admirable that we could reach the Moon with the technology available in the '60s. The truth must be known, especially now, because we are returning to the Moon. We must accept the facts and learn from our mistakes to evolve and do better next time.

Francisco Villate

February 2024

APOLLO 11
CONSPIRACY THEORIES

What do we think of when we hear about a conspiracy theory? That categorization has a different connotation for each person. Is it a waste of time? Is it a collection of various hypotheses with no scientific basis? Is it a mockery of people who lack logical thinking? Is something being hidden from us that we must unmask? Or was an event presented to us in a misleading manner? Is there new information that everyone should know but somebody will permanently hide from us?

There are a variety of conspiracy theories. Some of them turned out to be accurate, although they were thought not to be at the time. For example, the conspiracy theory said that for decades, tobacco companies hid the evidence that smoking is deadly. Interested parties attacked the conspiracy theory then, but it was true. And there are many more examples.

The problem with conspiracy theories is that around them, many branches grow that only contribute to disinformation and hide actual events. Sure, some conspiracy theories are outlandish from their inception. They seem like a social experiment to determine what percentage of the world's population will likely be convinced of something absurd. Others perhaps have a solid base, but then, whether unconsciously or deliberately, they are expanded with details and ramifications based on unreal and sometimes manipulated evidence, leading to misinformation. In the age of disinformation, how can we recognize the truth within that immense ocean of lies shown by the media or social networks? Disinformation has the effect of hiding the truth.

The methods we use to share information contribute to disinformation. For example, social groups like YouTube and similar ones allow us to upload content and receive advertising revenue. So, what is the real motivation for someone to communicate something on these media? Do they want to share something important? Do they want others to be educated and learn more? Do they want to reveal the truth to cultivate a more just and balanced society and thus build a better world? Or do they want to get rich by manipulating information and creating sensationalist news that attracts more visitors to their websites and videos?

Conspiracy theories attract many people who think they are tangible and should be made public. Conspiracy theories also cause others to reject them because of too much disinformation. Someone can quickly build a hypothesis with actual and manipulated facts, give it an attractive title, find a preferential place in the Conspiracy Theories Pavilion, and immediately guarantee the attention of many people and, at the same time, some economic income. Our information systems, social networks, the internet, traditional news media, etc., contribute significantly to disinformation. Today's technology allows disinformation to be more agile, simple, and forceful. For example, CGI tools enable any-one to create a simulation that looks authentic, such as a fabricated UFO landing in a park and ETs descending from the ship. Artificial Intelligence (AI) allows the creation of counterfeit representations of people, known or unknown, and sometimes famous, making it difficult to recognize what is real and what is not.

The biggest problem in our society today, which mainly affects young people, is that they base their knowledge on what they see on YouTube videos and TikTok or what they read on any social network instead of observing nature. They do not seek the truth in their outer, natural world or inside their inner world but give full credence to something they see on a social network.

What would happen if someday the Sun had a massive eruption that destroyed our satellites, leaving us without electricity and the internet? What behavior will those peo-ple who have been depending on that technology have, as if they were addicted to a drug? I am not trying to create a conspiracy theory or give a future prediction; I am just en-hancing the reality that, as a society, we are drifting away from reality as we become increasingly dependent on tech-nology.

It struck me as I read Billy Meier's statement that we will suffer from "phantasmagoria"[3] in the future due to techno-logical advances. He wrote many years ago about people being unable to distinguish reality from illusion. And we are already at that point. Some conspiracy theories go that way.

On the other hand, we can use technology to expand our knowledge because it allows us to be informed. We must think critically, analyze facts, and search for the truth. We should avoid the extremes of blindly accepting every conspiracy theory or dismissing them without making a logical analysis. That requires people to stop believing in gurus, masters, experts, researchers, official sources of information, politicians, and other leaders. Each person must be his leader and take charge of his life. And ride into the realm of reality and truth. The road is complex and full of obstacles of disinformation, manipulation through fear, and other techniques to suppress the power of our free decisions.

In the case of the Apollo program, especially Apollo 11, many people have created numerous conspiracy theories, some with a logical foundation, but others are baseless and contribute to misinformation. Let's examine the most well-known ones and verify their potential rationality.

Hypothesis 1:

The photos of the Apollo missions are false because we do not see the stars in the sky. There is no atmosphere on the Moon, and we must see them.

In all the photos and videos of the Apollo Program, like the Apollo 11 photo shown in Figure 1, we see a black night sky with no stars. We know that Earth's atmosphere makes the sky appear blue during the day, obscuring the stars. However, at night, the sky's lack of brightness allows us to see them clearly.

On the Moon, there is no atmosphere, so we should see the stars in the dark sky. Were these photos taken in a film studio, and someone forgot to paint bright spots on the ceiling?

Figure 1 - Apollo 11. NASA archive[5]

Because of my love for astronomy, I usually go out at night to look at the stars. Any astronomy enthusiast like me knows that seeing the stars in a bright environment is impossible. Our pupils need at least 4 minutes in the dark to dilate and be able to detect the faint light of the stars. If I am gathering with some friends at night and want to go to the patio to look at the stars, I see nothing in the sky when I come out. There are no stars. But after a while, and if the patio is dark, I can start to see them. But if the deck is luminous, I won't be able to see them because my pupils will remain closed due to the excess of light. Well, the same thing happens with cameras. They have a diaphragm, which, like a pupil, contracts so as not to receive excess light and overexpose the photos. And since the Moon's surface is bright, as the Sun fully illuminates the lunar soil, the camera's diaphragm is closed. It prevents the detection of the stars' dim light.

Suppose the astronauts land in a dark area on the Moon because the Sun does not illuminate it. In that case, we will see a very dark environment. The stars will be visible in the photos. Still, observing the equipment and instruments astronauts install will be challenging. Logically, NASA reached illuminated places so the astronauts could see what they were doing.

Hypothesis 2:

On the Moon, leaving footprints on the lunar soil is impossible because there is no atmosphere or humidity. A wet environment gives the granular soil the consistency that allows footprints to be imprinted.

Figure 2 - Footprint on Lunar soil during the Apollo 11 mission (NASA).

I have seen a test that some skeptics did in a vacuum bell. I'm glad to see someone try to check the facts for themselves. In the test, they placed a tray filled with sand and used a boot to imprint a footprint on it. Then, they duplicated the test, this time creating a vacuum inside the chamber. With the same tray and boot, they found that no mark was left when they dropped the boot onto the sand. The demonstration did not imprint a well-defined footprint like the ones in the NASA photos taken on the Moon.

Of course, if there is no air, the moisture in the soil disappears, and the sand grains no longer have cohesion. However, they did this test with Earth's sand, not lunar soil. What is the difference? On Earth, due to the effects of erosion caused by water and wind, the sand grains suffer wear that produces rounded edges particles. It causes them to lose cohesion. On the Moon, the soil is different. There is no erosion because there is no water or wind. Therefore, the lunar soil grains, which result from the fragmentation of the original rocks due to temperature variations, are not rounded. They have sharp edges, which makes lunar soil very corrosive.

The tests made by "The MythBusters"[4] are outstanding in verifying whether the Apollo program conspiracy theories have a scientific and accurate basis. They made an entire episode analyzing several of these theories and doing multiple experiments. They used the lunar soil that NASA gave them, which has these abrasive grains. They did the test inside a giant vacuum chamber. And indeed, by using lunar soil that has non-rounded grains, the footprint remains. Does that prove that the Apollo astronauts did get to the Moon? Not necessarily. It proves that making the footprints we see in the photos can be accomplished on the Moon or in a recording studio on Earth. Both hypotheses are valid. In this hypothesis, we cannot rule out that the astronauts have reached the Moon, nor that what we see in the photos is false.

Hypothesis 3:

In the NASA photos of the Apollo missions, we see the astronauts illuminated, even if they are in shadow, indicating additional lights to the one that simulates the Sun in a recording studio.

Figure 3 - Aldrin descends the Lunar Module stairway.
We see his suit illuminated. Photo NASA[5]

Figure 3 shows Aldrin descending the ladder of the Lunar Module. Armstrong took this photo, according to NASA records. We notice in this image that the astronaut is in shadow. The ladder is on the dark side of the Lunar Module. Those who defend this theory say a photographer took it in a recording studio because we see Aldrin illuminated. They attribute this to another cinematic light installed in a film set. So, in that studio, according to them, we have at least two light sources, a very strong one representing sunlight and a dimmer one that illuminates Aldrin so that he is visible in the photo.

If it is a fake photo, why install an additional light that could create other shadows and unmask the hoax?

The surrounding environment illuminates Aldrin. The ground is very bright from sunlight. Rays of light are reflected off the ground, reaching Aldrin's spacesuit and partially illuminating the dark side of the Lunar Module.

The MythBusters, in their demonstration[4], did a fascinating test. They used lunar soil in a mock-up with a scale model of the Lunar Module and an astronaut. They correctly noted that the test was conducted using lunar soil, which has the same albedo as on the Moon. Albedo is the percentage of light reflected by a material. For example, if the albedo of a soil is 40%, it reflects 40% of the light it receives. If the albedo is too low, the material will be very dark; if it is high, it will be very bright. In their test, they took pictures, and indeed, the surrounding soil reflected into the dark areas, giving that soft glow appearance that we see in the photos of the tests they did.

Again, does this mean that the Apollo astronauts did reach the Moon? Not necessarily, as both hypotheses are valid. The astronauts might have taken the photos on the

Moon's surface, or a photographer might have taken them in a movie studio. This hypothesis does not rule out lunar travel or prove that it did occur.

Hypothesis 4:

The shadows in several photos of the Apollo missions are not parallel. It indicates that there is more than one light in this place, not only one, to simulate the sun in a recording studio.

Figure 4 - NASA archive photo. The shadows are not parallel.

Skeptics say that, as we see in the photos, the shadows are not parallel, which means whoever took this picture used several lights, not just one simulating the Sun. However, if a photographer took this photo in a film studio on Earth, not on the Moon, we would see several shadows if he used several cinematic lamps. Still, there is only one on each object.

To understand what is happening, we must consider several factors. If the ground was flat and the astronauts had placed the objects vertically, the shadows cast by the Sun would not have been parallel in the photo. They would converge to a point near the horizon called the vanishing point. This effect is due to perspective. We can see this phenomenon by looking at the lines of a railroad projected toward the horizon. The railroad track may be straight, the ground flat, and the rails parallel. Still, we see them converging towards a vanishing point on the horizon.

What if the objects are not perpendicular to the ground but have some inclination? In that case, the projection of the shadow on the ground will also be tilted, not pointing towards the vanishing point. The instrument on the right, in the photo of Figure 4, close to the astronaut, was inclined to the left. In other images, we see it is not vertical and installed with some inclination. Its shadow does not point to the vanishing point marked with the letter "V" in the figure.

Additionally, we noticed that this photo is also tilted. We know that astronauts attached the camera to their chests; if the astronaut's posture was inclined, the image reflected that inclination. Here, we see a tilted horizon to the left. In many Apollo 11 photos and videos, we notice that the area of the Sea of Tranquility is relatively flat despite having some minor protrusions. The terrain is not sloping. The astronaut leans slightly to the right since the horizon falls to the left. That is the reason I draw the astronaut's centerline somewhat inclined.

The Lunar Module must be vertical and held upright on its four supports. We can estimate where the Lunar Module shadow casts by projecting the line that joins two supports. That line intersects the astronaut's shadow line at the vanishing point "V," close to the horizon.

The instrument close to the right of the Lunar Module is inclined to the right. We conclude this fact by comparing it with the central axis of the Lunar Module. For this reason, it does not cast shadows towards the vanishing point.

The conclusion is that the photo could have been taken on the Moon or in a large film set using a powerful and distant reflector to simulate the Sun. As with the previous hypotheses, both options are valid. Suppose someone were to fake the lunar environment in a video studio. In that case, they would avoid the awkwardness of using multiple lights.

Hypothesis 5:

The waving movement of the U.S. flag when astronauts installed it shows there is air around it. Thus, astronauts could not have taken this image on the Moon.

There are two questions about the flag. Why is it held in that position? Although the Moon has gravity, like Earth, objects should eventually fall without remaining in a horizontal static orientation. The second question is about its movement, which we see in the NASA video of the Apollo 11 mission. Other videos of other Apollo missions, such as

Apollo 17, also show the flag waving as if there were air. On the Moon, there is no atmosphere. The flag will not wave.

Figure 5 - Flag installed during the Apollo 11 Mission.
NASA archives photo[5]

On the first observation, it is clear after watching the videos that the flags have a horizontal support at the top, perpendicular to the flagpole. This support allows the flag to remain upright and not fall due to the effect of lunar gravity.

On the second observation, the flag waves as if it receives pressure from the air; we may know that what causes the resistance in a waving flag is not only the air but also the inertia. There may be no atmosphere, and inertia will still act, causing the flag's fabric to try to stay in its original position as it moves.

The MythBusters did a very good demonstration[4] in which they again used NASA's large vacuum chamber, where they set up a replica of the flag and waved it, the way we see it in the videos of the Apollo missions when the astronauts are driving it into the lunar soil. They did the demonstration with and without air inside the chamber. Interestingly, when you have air inside the enclosure, we see a dampening effect that causes the pendulum motion of the flag fabric to stop quickly. In contrast, when the operator creates a vacuum, not only does the flag undulate when shaken due to the effect of inertia, but it does not stop as quickly because there is no air to resist it. The MythBusters experimenters determine that what they see in a vacuum looks more like the NASA videos than what they observe when there is air. I agree.

This evidence leads me to conclude the first fact:

Fact #1: The flag movement in the NASA videos, such as Apollo 11 and Apollo 17, indicates that it occurred in an atmosphere-less environment, such as the Moon.

I base this conclusion on the videos we see in the NASA archives.

Hypothesis 6:

NASA could not have sent spacecraft to the Moon, like Apollo 11, because the technology at the time was not advanced enough to protect the astronauts. Cosmic rays would have killed them when crossing the Van Allen radiation belt in the Earth's magnetic protective field.

Figure 6 - NASA photo showing the Lunar Module returning to the service module on one of the Apollo missions.

There has been much debate about this. I conclude that NASA's technology allowed it to reach the Moon. NASA and Russia were able to land probes on the lunar surface. Landing a spacecraft carrying humans might have been more challenging at this time, but it was perfectly feasible.

Cosmic radiation was strong after leaving the Earth and crossing the Van Allen radiation belt. Some astronauts have died of cancer, without implying that it was due to the trips to the Moon. Here, I believe that NASA's explanation is valid, that the trips to the Moon were for a short time, only a few days, and that would not have caused the instantaneous death of the astronauts. I declare myself ignorant on this subject and would have to listen to a doctor's explanation of the effects of temporary cosmic radiation on the human body.

However, two pieces of evidence prove that the Apollo missions did reach the Moon. The first is proof of the reflection on mirrors installed by Apollo missions, which astronomers use to measure the exact distance between the Moon and the Earth. Observatories on the ground send a laser pulse to the Moon, and a signal back is received about two seconds later. By measuring the exact time it takes for the reflection to appear and knowing the speed of light, astronomers can measure the distance between the Earth and the Moon, which they calculate very accurately. Any astronomer from any country can use these mirrors in an observatory. If the mirrors do not exist there, astronomers from other nations would have evidence of a hoax perpetrated by NASA. However, there are no reports denying the presence of mirrors on the Moon.

The second and more significant piece of evidence is related to the photos taken by the Lunar Reconnaissance[6] space probe in 2009. This probe made a detailed survey of the lunar surface, and in one of its photos, we can see the landing site of Apollo 11 and the base of the Lunar Module, with its four supports.

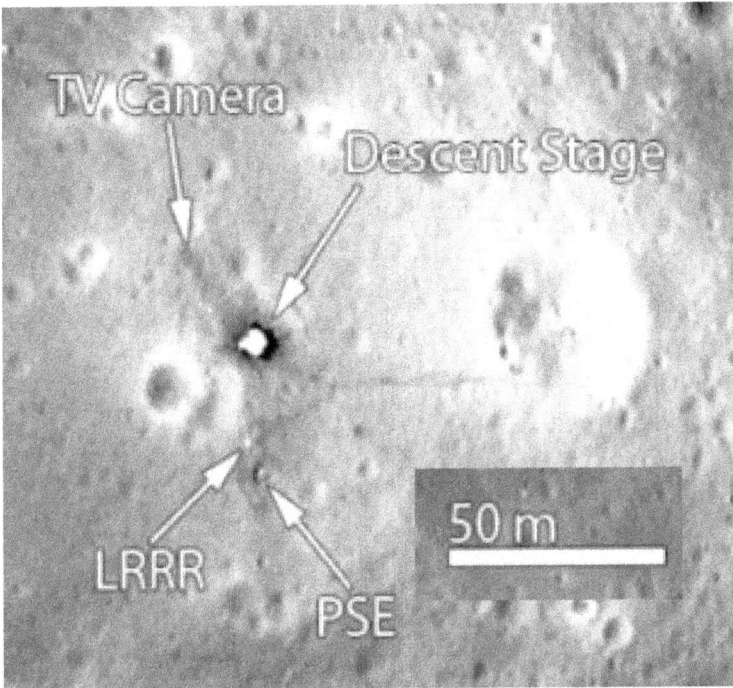

Figure 7 - Photo of the Apollo 11 landing site taken by the
Lunar Reconnaissance probe in 2009 (courtesy NASA).

Does this prove that Apollo 11 did land on the Moon?
Strictly speaking, not necessarily. Another mission to the
Moon could have installed this equipment. We can then
state the second fact in this analysis.

Fact #2: Several photos taken by the Lunar Reconnaissance probe in 2009 show Apollo 11 equipment installed on the Moon, indicating that some spacecraft put
it there.

Note that I do not state that Apollo 11 was the craft that installed the equipment there. This statement is in connection with later findings that indicate that the Apollo 11 astronauts could not have taken the photos we see of Apollo 11.

Some skeptics may declare that the photo shown here may be a fake. However, several photos from this probe are not false because other missions from countries like Russia, China, or India will take or have already taken pictures of this site. If they found evidence of nothing there, they would not remain silent about it. There is, indeed, equipment installed there.

Various skeptics put forward many other small details and hypotheses. I have included only the main ones, especially those related to Apollo 11.

NASA did indeed reach the Moon. However, as will soon be divulged, other facts show that Apollo 11 probably did not land on the Moon and that another space mission installed the equipment there.

WHO TOOK
THE APOLLO 11 PHOTOS?

We might wonder who took the photos of Apollo 11 that we see in the NASA archives and where they were taken from.

In my research, I have come to conclude:

Fact #3: The Sun is 6 to 7 degrees higher in the Apollo 11 Mission photos than it should be. The Apollo 11 astronauts could not have taken the Apollo 11 pictures on the Moon.

The Apollo 11 astronauts did not take photos during the "supposed" landing on the Moon. This fact implies that the Apollo 11 astronauts Neil Armstrong and Buzz Aldrin did not land on the Moon. And it means that we were all fooled. Somebody led us to believe they did, but it didn't happen as presented. In this chapter, we will summarize an investigation[7] that I started at the end of 2023.

Why have I concluded that the Apollo 11 astronauts did not take the pictures? The explanation is simple: the Sun is higher than should be seen in the NASA photo archives. The Sun is between 20 to 21 degrees above the lunar horizon in NASA's Apollo 11 pictures. However, it should be at an elevation between 14 and 15 degrees. When the Apollo 11 astronauts exited the lunar capsule to perform extravehicular activities, the Sun was at 14.2 degrees elevation. On the Moon, the Sun moves very slowly across the sky. It completes its cycle approximately every 29 days, unlike on Earth, where the Sun completes its cycle in 24 hours. This means that the Sun appears almost static on the Moon. When it supposedly took the Apollo 11 astronauts two hours to set up all their equipment, the Sun only moved one degree. When the astronauts returned inside the capsule, the elevation of the Sun was 15.2 degrees. So, it moved only one degree in the sky.

In reviewing the photos in the NASA records of Apollo 11, I see that the Sun is higher in all of the photos I have analyzed. On average, it is between 20 and 21 degrees in elevation. That equals a difference of 6 degrees. To put this in perspective, that 6-degree value is equivalent to 12 times the diameter of the Moon we see from Earth. Go out and look at the full Moon, and then imagine a distance in the sky of twelve times the diameter you see; that's 6 degrees of separation.

Is there an error in the value of 14 to 15 degrees elevation of the Sun? No, there is no error. On the NASA website, where we find the photos[5], they give an elevation of 14 to 15 degrees. Additionally, I checked using the Horizons[8] tool from the Jet Propulsion Laboratory (JPL). The Horizons tool is handy. It allows us to obtain ephemerides of the position of some celestial object, defining the place of observation and the time. With that tool, you can ask it to give you the elevation of the Sun (object to observe) at the Apollo 11 landing site (observation site) on the dates when the Apollo 11 mission was performed (moment in time when the astronauts were installing the equipment). The values I obtained were the same; when the Apollo 11 astronauts were supposed to take the photos during the extra-vehicular activity, the Sun was between 14 to 15 degrees elevation.

On the Moon, it takes 12 hours for the Sun to shift its position by 6 degrees. That means that the installation of the Apollo 11 equipment and the pictures taken by the astronauts must have occurred 12 hours later. However, the problem with that is that 12 hours after the extra-vehicular activities, the Apollo 11 astronauts were already lifting off from the lunar surface to meet up with Michael Collins, who was orbiting in the service module waiting for them to return.

When Apollo 11 landed on the Moon, I was 11 years old. It landed on July 20, 1969, and its astronauts embarked on their moonwalk late at night, according to the time in Colombia, South America, where I lived. I won't forget that day because it was Colombia's Independence Day. It was Sunday and a holiday. The next day, I had to go back to school. The live broadcast of the first moonwalk, which NASA presented on our black and white TV, was after 10 pm and until midnight. I remember watching it with my siblings and parents. And I remember us laughing at how the astronauts were hopping around like kangaroos while installing the equipment as if they were playing in an amusement park.

If the astronauts had installed the equipment 12 hours later, not when NASA records tell us, I should have been watching it at noon, from my school, not the night before, with my family. I remember very well that I was sleepy during my classes the following day, as I had gone to bed late the night before to observe this important event.

NASA recorded the time when astronauts took each photo, noted their communications, and logged the moments of each critical event during the mission. There was no doubt when it occurred.

The following table can be found in my research[7] and summarizes my findings from analyzing the elevation of the Sun in 12 photos.

In the table, I included the time astronauts took each photo in two ways: hours, minutes, and seconds from the start of the Apollo 11 mission, and in the form of date and time, given in Universal Time. I analyzed twelve photos, eleven of which they took during extra-vehicular activity, and the last one was taken from inside the capsule when the astronauts had already re-entered to get ready to sleep and then return from their trip. As seen in the table and all the photos, the Sun is higher, on average, by 6.8 degrees.

In my analysis, I used three methods, some by looking at the length of the shadow and others by using the reticle of the Apollo 11 cameras, which had been made to measure angles.

EVENT	Time from Takeoff (HH:MIN:SEC)	Date (UTC)	Sun Elevation during Apollo 11 (Deg)	Sun Elevation on Photos (Deg)	Difference on Sun elevation
Apollo 11 take-off	0	16/7/1969 13:32			
Lunar module lands on the Moon	102:45:00	20/7/1969 20:17			
Armstrong touch lunar ground	109:42:00	21/7/1969 03:14	14.2		
Photo AS11-40-5872	110:03:24	21/7/1969 03:35	14.4	24.4	10.0
Photo AS11-40-5873	110:03:24	21/7/1969 03:35	14.4	22.8	8.4
Photo AS11-40-5882	110:31:47	21/7/1969 04:03	14.6	20.6	6.0
Photo AS11-40-5884	110:31:47	21/7/1969 04:03	14.6	19	4.4
Photo AS11-40-5905	110:43:33	21/7/1969 04:15	14.8	25	10.2
Photo AS11-40-5931	110:55:49	21/7/1969 04:27	14.9	22.6	7.7
Photo AS11-40-5936	110:55:49	21/7/1969 04:27	14.9	20.2	5.3
Photo AS11-40-5946	111:04:56	21/7/1969 04:36	14.9	21	6.1
Photo AS11-40-5949	111:06:34	21/7/1969 04:38	15.0	20	5.0
Photo AS11-40-5961	111:11:31	21/7/1969 04:43	15.0	20.3	5.3
Photo AS11-40-5962	111:11:31	21/7/1969 04:43	15.0	21.6	6.6
EVA ended	111:39:13	21/7/1969 05:11	15.2		
Photo AS11-37-5466	112:20:56	21/7/1969 05:52	15.6	22	6.4
Compression of cabin / 5 hrs of sleep					
LM lunar lift-off ignition	124:22:01	21/7/1969 17:54	21.7		
		Averages		21.6	6.8

Table 1 - Expected versus actual elevation
found in twelve Apollo 11 photos.

High school students starting to explore trigonometry can do a straightforward experiment to determine the elevation of the Sun. They can take a bar or pole, measure its length, go out to the schoolyard, place the stick vertically on a flat, horizontal ground, and measure the length of the shadow. The students can calculate the elevation angle using the lengths of the shadow and the pole. To find this angle, they must divide the pole's length by the shadow's length and apply the inverse tangent function. Students can use bars of different sizes simultaneously, and the angle value will be the same. Or anyone can take a picture in a front view of the triangle formed by the pole and its shadow and measure those dimensions in any unit (millimeters, inches, etc.) on the photo. The resulting angle value will be the same.

You can do the same thing with some of the Apollo 11 photos. Although the terrain might not be completely smooth or horizontal, this method can give you a reasonable estimate. You may expect the values to be in the same range of 19 to 23 degrees, not 14 to 15 degrees. When I first used this method and studied some photos, I expected to obtain values centered around 14 or 15 degrees. That is, to obtain, for example, values of 13 to 16 degrees. But no, I did not get that value in the first photo I analyzed. I got 19 degrees. Then, I looked at other pictures and found values above 20 degrees. What was going on? Why such high figures?

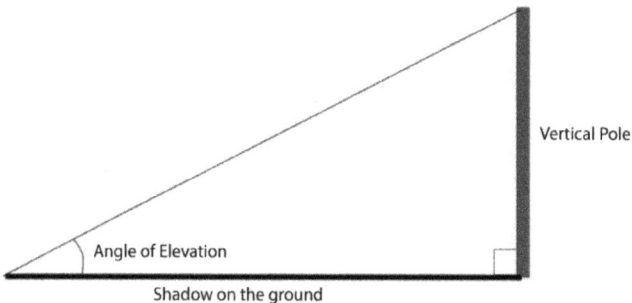

Figure 8 - Sun elevation angle from the lengths of a pole and its shadow.

The following (Figure 9) is a photo that I analyzed. It is the photograph cataloged as AS11-40-5884, taken from the NASA page. It is a high-resolution photo, and you can enlarge it to see the shadow it makes on the lunar soil in more detail. It is the TV camera that the astronauts placed on a tripod. At the top, you can see the camera. On the ground and to the left is the shadow. By enlarging the photo and measuring the height of the camera above the lunar ground and the length of the shadow, I obtained 19 degrees of elevation using the method already indicated. To do this, I measured the height of the top of the camera above the ground as 57 millimeters on my computer screen. The length of the shadow I measured was 166 millimeters. Dividing both values and with the inverse tangent function, I got 19 degrees. But the angle should be 14.6 degrees, not 19 degrees. If it were 14.6 degrees, the shadow would be 32% longer than we see here. That is a lot.

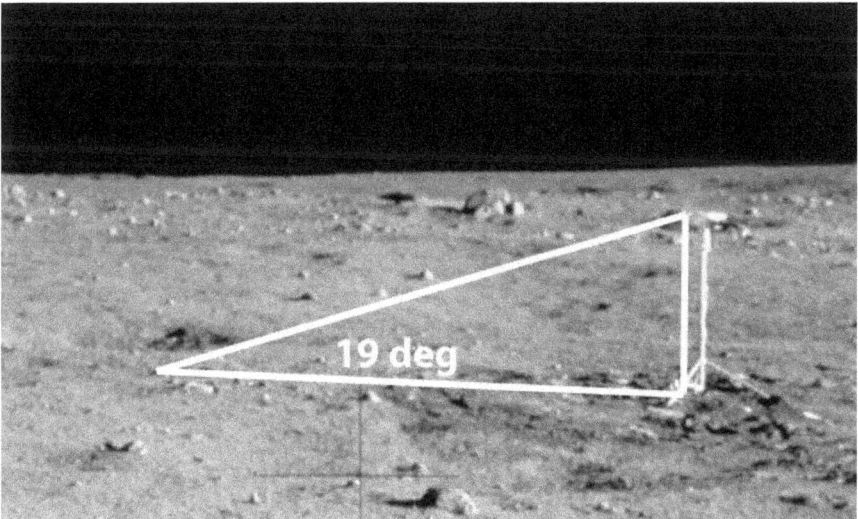

Figure 9 - Photo AS11-40-5884 of Apollo 11.
The elevation angle is 19 degrees.

This method works well if we have that triangle formed by the vertical pole and the shadow in a frontal way. The shadow must run from left to right or right to left in the photo, not backward or forward. If the shadow appears in perspective, making this calculation directly is no longer accurate. So, with this method, we can analyze only a few photos that show the shadow and the pole (or a vertical line) in a front view. Many are in perspective. Another difficulty is that in most of the photos, the shadow goes out of the photo's field of view, and we cannot find where it ends or measure it. However, it gives us a reasonable first estimate with a sample of a few photos.

Another method used was to analyze objects whose shadow appears in perspective. That is to say, the Sun casts the shadow backward or forward. In this case, the technique consists of rotating the triangle formed by the vertical pole and the shadow to locate it frontally. The following figure, from the NASA photo, AS11-40-5905, showing the U.S. flag, has the triangle A-B-C, which, when rotated, yields the triangle A-B-D. From a top view of the place where the flag was, we can imagine a circle on which the indicated triangle rotates. In the photo, we will see that circle as an ellipse because it is in perspective.

By looking at this photo in detail and observing the shadow of the vertical flag pole on the lunar soil, we can conclude that the astronauts set the flag on a small mound about 20 centimeters above the surrounding ground. Then, the vertical measurement must be made to the base of that mound, not to its upper surface. When examining another photo of this flag with an astronaut near it, some skeptics noticed that the pole's shadow had disappeared and considered the photo a fraud. It disappears because it is behind the mound. These photos are real, but were they taken on the Moon? Did the Apollo 11 astronauts take them?

In this method, drawing an ellipse of correct dimensions is essential. That is, the relationship between the minor and major axes of the ellipse must be correct. My research describes how we can draw this ellipse with the correct dimensions.

In this photo, AS11-40-5905, the Sun should be 14.8 degrees above the lunar horizon, but I found 25 degrees. That is a big difference.

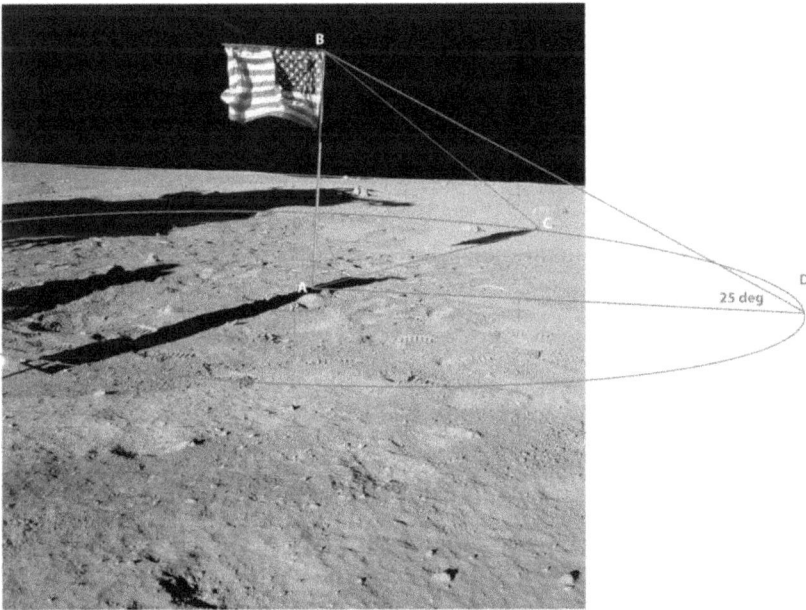

Figure 10 - Analysis of the photo AS11-40-5905. The triangle shows the elevation of the Sun when we rotate it into a front view.

The methods given so far seem approximate. But in reality, they provide values that, on average, can tell us the Sun's actual position in the photos I have analyzed. The images show an elevation higher than 14 or 15 degrees.

Why?

The following method, which I performed in two ways, is much more accurate and is the best of all, giving us more reliability in the results obtained. The method uses the characteristics of the camera utilized by the astronauts in the Apollo missions. NASA used a medium format camera, a Hasselblad 500EL[9], which employed a Zeiss lens specially designed for NASA. The camera had an intermediate glass, a "Réseau Plate"[10], with a five-by-five crosshair pattern shaped like a "+." The central mark is longer than the others to differentiate them. The purpose of this grid was to accurately measure the horizontal and vertical angles of objects in the photos. So, we cannot expect incorrect results when measuring angles based on this reticle, but rather very accurate values due to the high quality of the cameras.

That camera can use several lenses, including 60mm, 80mm, and 500mm. The astronauts used the 60 mm lens for their photos during the extravehicular activity[11].

Figure 11 - Hasselblad 500EL camera, which the Apollo mission used to take pictures on the Moon (Credit: Hasselblad).

In addition, and to give astronauts greater freedom of movement, these cameras could be attached to the chest of the spacesuit. This configuration allowed the astronauts to move nimbly, hands-free, and not worry about holding a camera in their hands or having it fall to the ground. In the videos, we see the astronauts using the camera strapped to their chests most of the time, and at other times, they carry it in their hands.

Having the camera fixed on the chest makes some photos appear tilted. The astronaut did not necessarily have a vertical body posture when taking the pictures. In addition, this feature allows us to know precisely what height or where the camera was in each photo.

Figure 12 - The location where the camera is attached to the chest of the Apollo 11 astronaut. Aldrin photographed Armstrong.

The method of measuring the Sun's elevation angle using this reticle pattern has two options. One is to look for the direct location of the Sun in the photos, even if it is slightly out of the field of view, and the other is to look at the point opposite the Sun. Both allow us to determine the elevation very accurately.

The next photo, AS11-40-5936, taken by Aldrin, shows the lunar horizon and several solar rays and indicates that the Sun was very close to the upper edge of the photo. We also clearly see the "+" markings, which, according to NASA, are 10.3 degrees apart, as indicated in the Réseau Plate details[10].

Figure 13 - Photo AS11-40-5936.

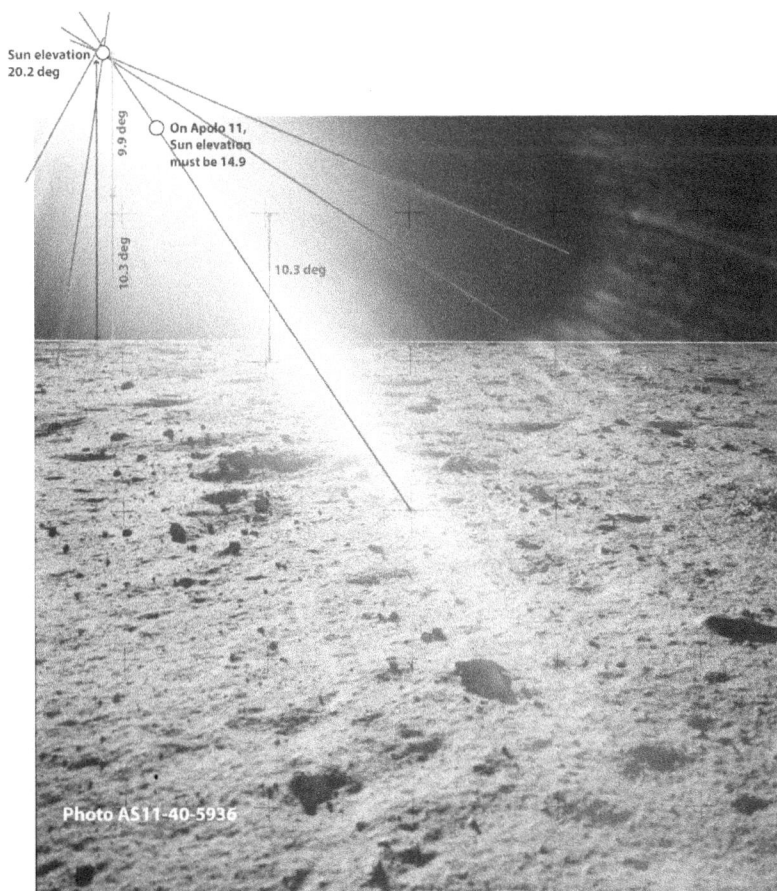

Figure 14 - Photo AS11-40-5936 analysis.

In the same photo, if we project the rays that appear in the image, they converge where the Sun is. And we have many rays that allow us to locate it very precisely. In figure 14, we show those rays and the location of the Sun, which, when we measure its elevation, gives us 20.2 degrees, but it should be 14.9. Why so high? It is well above where it should be if this photo were taken by Aldrin when he was on the Lunar surface. At the time when Aldrin shot this photo, he must have already been flying in the capsule, along with Armstrong, not still walking on the Moon and taking pictures.

We can do a reverse analysis on this same photo. We can determine where the Sun should have been in the picture if it were at 14.9 degrees, placing it just inside the field of view. But we don't see it in the photo. Not only should we observe it in the image, but being inside the photograph would make it veiled and totally white because the Sun's rays would have entered the camera and damaged the photo's negative.

In the second part of the method for determining the elevation of the Sun, we find the point opposite the Sun. That is 180 degrees in the opposing direction. Figure 15 shows a schematic to explain what this method is all about. If we draw a line from the Sun passing through the camera lens, this line points to the opposite position. If, for example, the astronaut's shadow appears on the lunar soil in a photo, and we locate the place where the camera was between that shadow, we can measure its separation from the lunar horizon. This dip angle of the direction of the point opposite the Sun is the same as the elevation angle of the Sun. This method allows us to measure the Sun's elevation in a photo without risking damage to the camera or our eyes from direct exposure. We can do this by finding the point opposite the Sun.

This method is effective regardless of whether the astronaut is positioned in front of a depression or on a mound of lunar soil. The terrain profile does not affect the angle measurement; it is only necessary to be able to observe the horizon in the same photo.

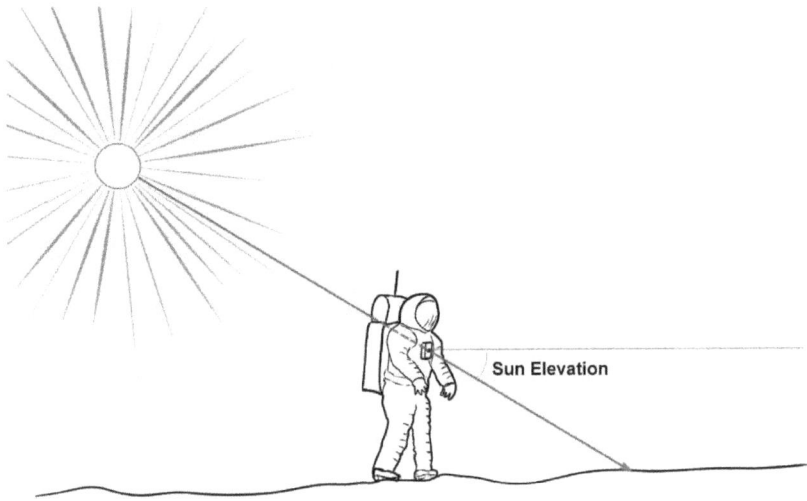

Figure 15 - The elevation of the Sun is the same as the dip of the opposite point.

Three photos meet these characteristics. I am going to show one of them, photo AS11-40-5961. I presented this photo in the previous chapter. Here, I use it to take the corresponding measurements.

On my computer screen, I measured several distances in millimeters, like the separation between the marks of the photo reticle, which we know to be 10.3 degrees[10] (the camera using the 60mm lens), resulting in 80 millimeters. A little square indicates the camera's location in the astronaut's shadow in figure 16. We know he carries the camera on his chest. And from there to the horizon, we have a separation of 158 millimeters. With these proportions, it is easy to calculate that the dip angle of the point opposite the Sun was 20.3 degrees. However, the Sun's elevation angle must be 15 degrees if the Apollo 11 astronauts really took it.

33

Figure 16 - Photo AS11-40-5961 analysis.

The round dot above the astronaut's head represents where the camera's projection should have been on the lunar soil if the Sun had been at 15 degrees elevation. Suppose this astronaut took the camera in his hands and raised it over his head, which we do not believe he did. So why don't we see the shadows of the camera, nor the astronaut's hands at that point?

Like the others, this photo clearly shows us one reality: The astronauts of the Apollo 11 mission could not have taken the Apollo 11 photos we find in NASA's records.

Who took them? Was it another secret Apollo mission? Did somebody take the pictures in a recording studio that simulated the lunar landscape but installed the reflector too high? Who mounted the equipment we see in the Sea of Tranquility in the *Lunar Reconnaissance* probe photos?

Finally, this is the sequence of events during the Apollo 11 mission, when the astronauts landed on the Moon (Figure 17).

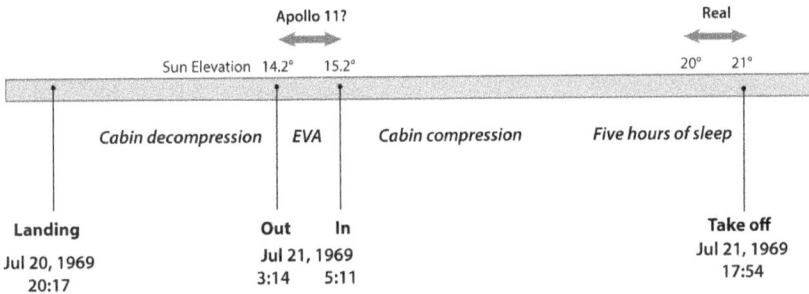

Figure 17 - Events during Apollo 11's stay on the Moon.

The Lunar Module landed on the Moon on July 20, 20:17 Universal Time (UT). Over the next seven hours, the astronauts prepared the installation equipment, put on their spacesuits, removed all oxygen from the cabin by decompressing it, and opened the door to go outside. On July 21, 3:14 UT, Armstrong descended for the first time and stepped onto lunar soil. Aldrin followed. The two set up all equipment during extra-vehicular activity (EVA) in the next two hours. The Sun was between 14.2 and 15.2 degrees in elevation. They returned inside the cabin at 5:11 UT. They closed the door and re-created compression inside the module cabin, waiting until the oxygen filled it and had adequate pressure. They removed their spacesuits. According to NASA, after this, they slept for five hours. At 17:54, they lifted off from the Moon to rejoin Collins, who was

35

waiting for them up in the service module to return to Earth.

The angle I have found from the Apollo 11 photos is between 20 and 21 degrees elevation or a little more than that, and this occurs as they are already lifting off. They couldn't have gone back outside because that would require losing all the oxygen in the cabin again. Additionally, after returning, they would need to wait several hours for the Lunar Module cabin to regain compression for liftoff and then take off several hours late. If something different happened during the mission, why didn't NASA say so? Did they go back out? That's impossible, as the live TV feed, which I watched around midnight on July 20, 1969, clearly showed them installing the equipment, and it doesn't make sense for them to have dismantled and then reinstalled everything just to take the final pictures. And it doesn't sound logical that NASA, who keeps the exact record of hours, minutes, and seconds of every photo, event, and dialogue that occurs, would have made a mistake. What we saw on television does not indicate a change of plans.

This evidence shows that the Apollo 11 astronauts couldn't have taken the pictures we see in NASA's records. Did they pretend to land on the Moon, only for subsequent Apollo missions to accomplish the task? Were the photos we see taken by other astronauts on another secret mission in the same place but at a different time? Or were those photos taken in a film studio? Or do the images mix photos taken in a video set with pictures from another mission?

I will call that hypothetical Apollo mission "Apollo Y," which installed the Apollo 11 equipment we see in Lunar Reconnaissance probe photos.

THE BIZARRE
APOLLO 11 VIDEO

NASA has posted some videos on its YouTube channel, two of them from Apollo 11. One video captures the Lunar Module's descent, recorded with a 16mm film camera mounted inside the Lunar Module, positioned in front of the right window, and aimed downward. That camera recorded the descent onto the lunar surface. That video and the computer simulation are effectively compared in a YouTube video[12], alongside images from the Lunar Reconnaissance Orbiter.

The second video[15] is about the extravehicular activity when Armstrong and Aldrin descended the ladder and installed all the equipment. This recording is supposed to be the same video we saw in 1969 in a TV direct transmission from the Moon, but it is not. And the footage we see today is a peculiar recording that gives us more clues about what transpired during the Apollo 11 mission. This video is different from the original, as the original version, which was registered and contained within several magnetic tapes, was lost. In 2006, NASA admitted that the original tapes were lost and that they likely had been overwritten during subsequent Apollo missions[14]. So, where did the video NASA posted on its YouTube channel come from? NASA says it is an enhanced recording of different copies from different sources, one of which comes from the CBS News archives[15]. Interestingly, another copy has recently surfaced, which is part of an exhibition in a museum in Florida, which is perhaps the same restoration created by NASA. Either way, we cannot compare the original version with the official version on NASA's YouTube channel because the original was lost.

Fact #4: The actual video of the Apollo 11 mission with its activity on the Lunar surface is not the original. It is a recent copy. The original no longer exists.

Losing the original tapes and using an unofficial recording has logically awakened and supported many conspiracy theories about Apollo 11. Additionally, looking at the video in detail is enough to realize the first part was not taken on the Moon, not in the Sea of Tranquility. The video has two parts, maybe three, which makes it look like a sequence of the same video. Still, whoever made these recordings did it in different places and with different cameras. I will detail that in this chapter.

TELEVISION SIGNAL TRANSMISSION FROM THE MOON

And how was a television signal transmitted from the Moon? There is a lot of technical data about television cameras, their evolution, and how Apollo 11 sent the signal from the Moon. In simple terms, the video we see in the NASA record, which is not the original, was transmitted with a camera that gives a black-and-white image in a simple format that made the transmission possible. When it reached the ground stations, the electronic equipment converted it to the format used by the television of that time. A transmission system developed by NASA, which they currently use to transmit voice, data, and images from spacecraft, employs a microwave system called "Unified S-Band" [16]. The Lunar Module had a small antenna on its top that could transmit such a signal as long as large antennas on Earth received it. The spacecraft carried an additional, portable, umbrella-like antenna, which we see on other Apollo missions. Still, the Apollo 11 astronauts did not install it.

Figure 18 - Goldstone (California) Station is on the left, and Parkes (Australia) Station is on the right. Both measured 64 meters in diameter.

At the time of transmission on July 20, 1969, around midnight Eastern Standard Time, the Moon was not visible in most of the United States; only Goldstone Station in California could receive the first few minutes of transmission. Then, the Australian station at Parkes received the signal during the remainder of the extra-vehicular activity. These two stations that received the television signal from the Moon had large antennas, 64 meters in diameter, which allowed them to receive the signal from the small antenna of the Lunar Module.

Figure 19 - Antennas for transmitting television signals from the Moon. On the left is the Apollo 12 external antenna. The small Apollo 11 antenna on the Lunar Module roof is on the right.

Inside the Lunar Module, they carried a special antenna to transmit the signal via television, consisting of thousands of meters of gold wire, which looks like an umbrella[17]. Apollo missions used these antennas. So why didn't the Apollo 11 astronauts install the higher quality external antenna after they descended? They wanted to transmit Armstrong's descent to the lunar surface for the first time. The small antenna might be enough. But why didn't they install the big one they had on board? It should have improved the signal. The image of the transmission from the Moon could have been in better resolution. We know that the astronauts spent several hours on Earth conducting simulations to prepare for the sequence of installing each piece of equipment on the Moon. Wasn't the installation of this antenna already planned as part of the activities they intended to perform?

Fact #5: The Apollo 11 astronauts carried an antenna to transmit a better-quality television signal but decided not to install it. That led to receiving a poor-quality signal sent from the Moon.

You can compare TV signals from other Apollo missions with those from Apollo 11 to see the significant differences.

TILTED IMAGE

Figure 20 - Detail of the TV camera placed upside down on an external support.

NASA explains that the camera was mounted outside the Lunar Module and captured the first images of Armstrong descending the ladder to the lunar soil. In Figure 20, we see a picture of the camera on an external support, and we notice that it is tilted. NASA explained that this inclination caused the image of the lunar horizon in the Apollo 11 video to be inclined 11 degrees. We can compare it with the Apollo 12 video when the astronauts descended to the lunar soil (Figure 21). They also carried the TV camera on the external platform. In the Apollo 12 video[18], the horizon is also inclined. When they took the camera and inverted it, the image was upside down. The Apollo 12 video shows remarkable differences from what we see in the Apollo 11 video (Figure 22).

Figure 21 - Comparison with this image of an Apollo 12 astronaut. The horizon is tilted, as is the astronaut. The original image has been inverted in this figure.

An inverted camera transmitted the first moments of the Apollo 11 moonwalk. The signal operators must have inverted it to correct it. In its first part, the NASA video happened when the camera is on this external support of the spacecraft. The second part occurred when Armstrong detached the camera, held it, and installed it on a tripod a few meters away from the spacecraft.

The first part of the video is quite odd. It was not recorded in the same place as the second part or with the same camera. Let's analyze the first part of the video.

I have extracted several shots directly from the NASA video on YouTube, as shown in the following figures, and you can see each marked with the exact time they appear. The first surprising thing is that if the camera was tilted. One would expect to see the astronauts walking at an angle. The following figure shows a shot twelve minutes into the video. We watch the astronaut upright, and the horizon tilted 11 degrees. Why is the astronaut not inclined as well? I have drawn a profile of what astronauts should look like in Figure 22. Their vertical axis should also be tilted to the right 11 degrees. I invite you to watch the video, see how Armstrong and Aldrin walk and move near the spacecraft, and notice that their axis of symmetry is not inclined. It is vertical.

In the case of Apollo 12 (Figure 21), we also see the tilted horizon in another NASA video[18]. The astronaut is inclined, different from what we watch in Apollo 11. And it is evident in the Apollo 12 video when the camera is inverted, but this camera inversion is not observed in the Apollo 11 video. NASA documentation explains that the ground station operators receiving the Moon's signal could invert it back if they initially received an inverted image. So, the signal operators would have viewed the inverted image, and then they must have corrected it. All viewers world-wide would see the inversion events, but it did not happen. I do not remember seeing that as a child on July 20, 1969.

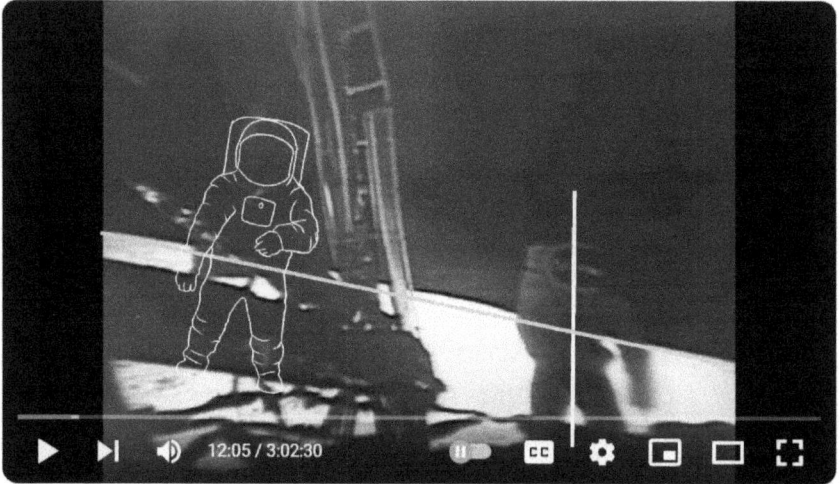

Figure 22 - Armstrong walks near the Lunar Module.
The horizon is tilted, but he is not.

The Apollo 11 astronauts removed the upside-down camera and held it by the handle. The image must be inverted at that moment, as seen in the Apollo 12 video. However, in the Apollo 11 video, we never see the image inverted.

Fact #6: Even though the TV camera starts upside down and rights itself when taken by the astronauts, we do not see in the NASA video that the Apollo 11 image is inverted at any time.

Who took the Apollo 11 video, and where did it happen? The Sea of Tranquility, although it has small bumps, is horizontal. Why do we see the astronauts walking along it on what looks like a hill sloping 11 degrees? And that's not a small bump; that's the far horizon. The second part of the video, when Armstrong moved the camera away from the spacecraft and put it on a tripod, shows us a relatively flat, horizontal landscape. It looks like a different place.

ANOTHER CAMERA?

Another striking aspect has to do with the camera lens. In the second part of the video, we see an internal reflection effect of the lens that we did not see in the first part. This effect produced by camera lenses, often referred to as lens flare, results in a ghost image of a bright object being projected opposite to the center of the image and inverted. Additionally, it is of lower luminous intensity. The following figure shows a photo my sister took with her cell phone during a solar eclipse. The Sun appears distorted on the upper right because it is very bright. That image of the Sun is reflected internally on the surfaces of the different lenses of the camera lens. It produces a ghost image, which appears at the lower left. This method is one way to take pictures of eclipses with a cell phone camera without using special filters. Please do not repeat this experiment because I do not guarantee that it will not burn your camera sensor.

The NASA Apollo 11 video shows this effect in the second part. When observing it, we notice some glows, like ghosts, moving across the image. If you look closely, you will see that they move symmetrically and inverted to the center of the picture. In the following figure, I show an example in a video shot.

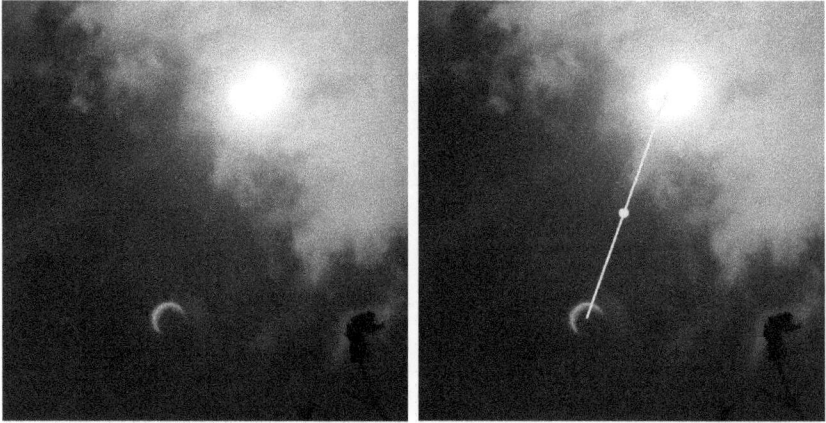

Figure 23 - Effect of the reflection in the lens. During a partial eclipse, the Sun is at the upper right and creates a ghost image opposite the center of the picture. The image looks like a crescent moon, but it is a reflection of the Sun.

In Figure 24, we see two lens reflections: one near the astronaut, which is the reflection of the left edge of his spacesuit, and the other seen as a diagonal bar to the right, which is the reflection of one of the Lunar Module base supports. These are obvious in the video as they move in sync with the original image. However, they are much more noticeable in the second part of the video than in the first. When Armstrong and Aldrin descend and walk near the spacecraft, there are moments when the Sun fully illuminates their spacesuit. Still, we do not see reflections. That gives the impression that it's another camera, at least with another lens. It may be two different cameras in different places.

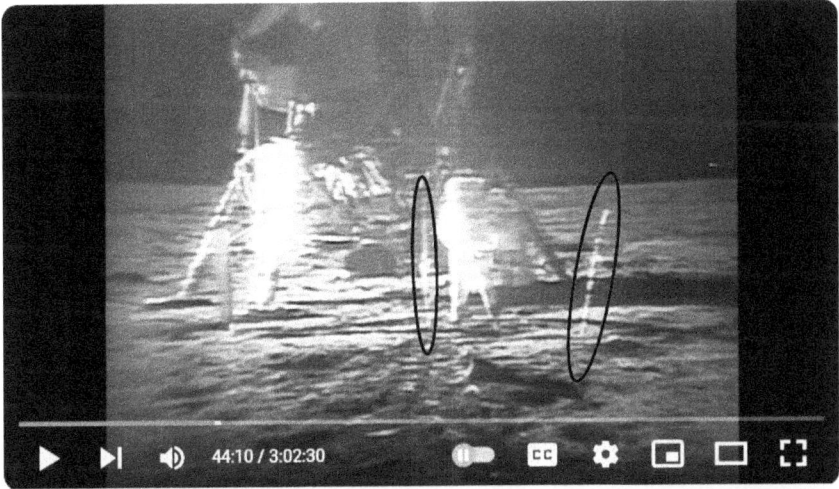

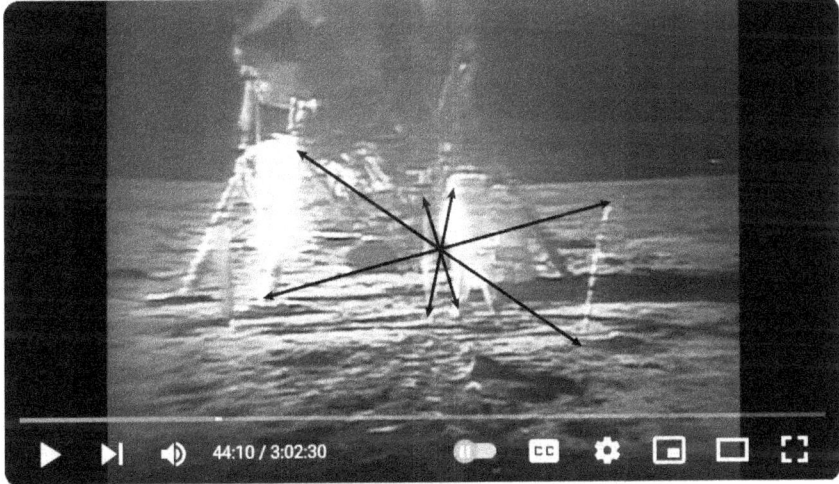

Figure 24 - Lens reflection effect. Two glows that appear
in the video are reflections on the lens.

Note the position of the lunar horizon in the figure. It looks horizontal, not at an 11-degree tilt. The astronauts are not on a hill but on a plain, just as the Sea of Tranquility is.

The NASA video implies this is a continuous shot with the same camera, but it is not. At one point, Armstrong picks up the camera and moves to another location; at times, the image is entirely white. That is where the cut between the two takes is.

Something curious is that there is an intermediate portion of the video where the image improves substantially; it is for a short time and happens when they read a plate installed in the Lunar Module. For a few brief moments, we see the effect of the reflection on the lens. The description of the technical part of the lunar transmission[16] says that at that moment, the image improved because the Australian antenna would start transmitting. It was in a better position than the one in California. That sounds logical and coincides with the brief improvement of the transmission. What is not logical is why the broadcast is later worse when Armstrong puts the camera on the tripod away from the spacecraft. We see the Lunar Module and the raising of the flag in inferior video quality.

SUN'S ELEVATION

Let us now analyze the elevation of the Sun in the video, to verify if there is any correlation with the results we found in the previous chapter, where we concluded that it was very high. Figure 25 shows the first group of frames taken from the first part of the video.

We see the Sun cast shadows from left to right. Thus, we can measure the angle because the triangle formed is in front view, in its actual dimension. Each frame shows the time in the NASA video. The astronaut was in a position in which we could see his shadow. The values obtained are 32, 34, and 35 degrees, which is much higher than the 14 degrees that the Sun should have in elevation and higher than the 20 degrees that the Apollo 11 photos show.

Figure 25 - Elevation angle of the Sun in the first part of the video.

The shadows suggest a flat terrain, yet the far horizon is tilted. It gives the impression that the lunar horizon is fake, perhaps modified with a tool like After Effects to make it look sloped. Somebody could have filmed the first part in a film studio, where they forgot to tilt the camera. In post-production, they decided to edit the video and show the horizon tilted as it must be. However, editing the tilt of the horizon does not change the vertical position of the astronauts.

Let's now check the Sun's elevation in the lunar environment shown in the second part of the video. Will we find the value of 14 to 15 degrees consistent with the date and time the Apollo 11 astronauts were on the Moon's surface? Or will we obtain 20 to 21 degrees like the actual elevation in Apollo 11 photos?

This analysis can be done directly from the video since the Sun casts shadows to the right, and the triangle that we observe in front view displays the actual elevation of the Sun. Determining where the astronaut's shadow ends in these low-resolution images is difficult because of the poor-quality video. In the photos shown here, it can be challenging. However, relying on the video is straightforward because we can see how the astronaut's shadow moves simultaneously as the astronaut walks on the lunar soil—that way, we can determine where the shadow ends. Then, we measure the sides of the triangle, apply the inverse function of the tangent, and get the Sun's elevation angle. Figure 26 shows two results.

Again, we find the same values as the Apollo 11 photos, which we know were not taken by the Apollo 11 astronauts. We get 21.3 and 19 degrees. In the second part of the NASA video, we observe astronauts moving in a flat environment, with no atmosphere and low gravity, in a place that resembles the Sea of Tranquility on the Moon. Notably, the Sun appears very high, suggesting a time that does not align with when the Apollo 11 astronauts were there. Where is this place? Is this the right site, but at another time? Is this the mystery Apollo "Y" mission that installed the equipment in exchange for Apollo 11? Is the video from that secretive mission? Who are the astronauts walking around there?

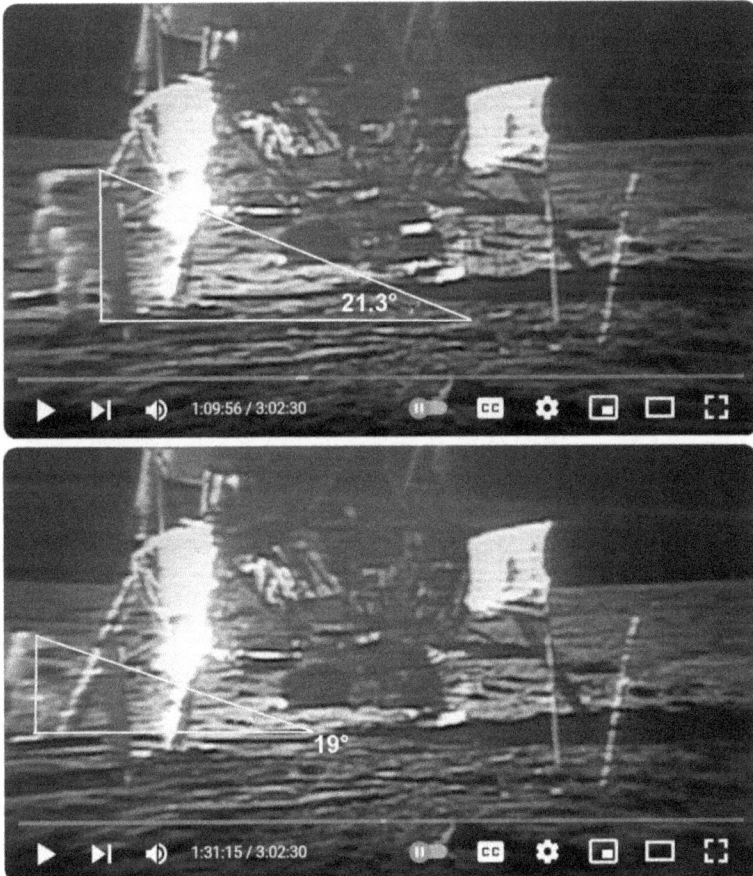

Figure 26 - Elevation of the Sun in the second part of the video.
We obtain 21.3 and 19 degrees.

Here to conclude:

Fact #7: The NASA Apollo 11 video shows two very dissimilar environments, indicating that it comes from two different shots, taken in separate locations and with different lenses. The elevation of the Sun in the first part is too high, about 34 degrees. In comparison, the second part of the video shows the height of the Sun at about 20 degrees, similar to that of the NASA photos. Neither matches the expected value of 14 degrees from the Apollo 11 Mission.

PENDULUM MOTION

The astronauts wear a buckled ribbon on their space suit, which bifurcates at the lower end. We find that ribbon in Figure 12, where the astronaut poses for a photo. The ribbon detaches from his chest, just below the camera mount, and is about 40 to 50 centimeters long.

In the video of the astronauts walking on the Moon and installing all the equipment, we notice how that ribbon moves in a pendulum motion. The period of that pendulum's movement depends on the place's gravity. That's a scientific method for measuring the acceleration of gravity in different locations, such as on the Earth or the Moon. The higher the gravity, the faster its pendulum motion. That is, it will move faster on Earth than on the Moon.

Using a video editing tool, I analyzed frame by frame several moments in which that ribbon swung like a pendulum and measured its period. The period refers to the time it takes the suit's ribbon to swing from one end to the other and back to its original position. To ensure accuracy, I looked for moments when the astronaut was standing still, not walking.

In the first part of the video, you notice that the movement of the tape fades quickly. This effect happens in an environment with an atmosphere, as the air gradually stops the pendulum motion. In the second part of the video, the movement remains for a long time, indicating no air to stop it. For example, when "Aldrin" poses for the picture, saluting the flag just hoisted on lunar soil, we see the ribbon pass through the space between his left hand and his body several times.

In the following table, I indicate the moments at which I measured the pendulum motion, the corresponding period of the pendulum, and the equivalent length if the astronaut were on the Moon (low gravity) or Earth (higher gravity). I used the well-known pendulum formula, familiar to any high school student.

Time Stamp (mm:ss:ff)	Period (Seconds)	Average Period (Seconds)	Lenght on Moon (centimeters)	Lenght on Earth (centimeters)
10:01:13				
	1.56			
10:03:00		1.58	10	62
	1.60			
10:04:18				

Table 2 - First part of the video. Measurement of the pendulum movement of the belt.

The table shows the time I measured the pendulum periods in the first column, indicated as minutes: seconds: frames. The number of frames is 30 (0 to 29) because the video comes in 30 fps. The time difference means the period of the pendulum movement. I took two measurements and averaged them. Then, I used the pendulum formula to calculate how long the equivalent pendulum should be.

If the astronauts had recorded the video on the Moon, the ribbon hanging from Armstrong's suit should have measured 10 centimeters. It is about 40 to 50 centimeters, so it is absurd. And if they had recorded on Earth, that ribbon should be 62 centimeters, slightly longer than it is. Why doesn't any measurement match the actual length of that tape? What is more likely, and might explain it, is that the cameraman recorded it on Earth's environment and later edited the video to make it look slower, to give the impression the astronauts were on the Moon. There is a 60% delay in the movements. The idea that an astronaut recorded this video on the Moon and a video editor accelerated the astronauts' activities for some unknown reason seems illogical. If this were the case, the astronauts would appear to be moving unnaturally fast. In reality, we see them moving slowly.

Now, let's analyze the pendulum motion in the second part of the video when Aldrin poses for the photo next to the flag.

The results are consistent with a lunar environment (Table 3). The pendulum has an effective length of 30 centimeters. It makes sense that the effective length is shorter than the actual length, which ranges from about 40 to 50 centimeters because the mass of the ribbon is distributed along its entire length. Obtaining 30 centimeters is correct. If it had been on Earth, the ribbon would have a length of almost two meters, which is absurd. That leads to the following conclusion:

Fact #8: The pendulum motion of the tape hanging from the astronauts' suits, specifically its oscillation period and how it is damped, indicates that the first part of the video was recorded in a terrestrial environment with an atmosphere, and the second part in a lunar environment with low gravity, and no atmosphere.

Time Stamp (mm:ss:ff)	Period (Seconds)	Average Period (Seconds)	Lenght on Moon (centimeters)	Lenght on Earth (centimeters)
51:09:04				
	2.8			
51:11:28		2.7	30	181
	2.6			
51:14:17				

Table 3 - Second part of the video. Measurement of the pendulum movement of the belt.

That means we see a combination of two videos: the first part was not taken on the Moon in the Sea of Tranquility but on Earth, most likely in a film set, and the second part was recorded on the Moon. That does not mean that Armstrong was indeed on the Moon's surface, as we have already shown that the Apollo 11 astronauts could not have taken the Apollo 11 photos we see in NASA records. So, who are the astronauts posing as Armstrong and Aldrin? They are astronauts from the mysterious Apollo "Y" mission.

KICKING THE LUNAR SOIL

I have analyzed in detail the movement of the astronauts and how they hit the lunar soil grains with their boots in the second part of the video, where we see a landscape like the plains of the Sea of Tranquility. I found something very revealing. In the 49th minute, the grains that one astronaut disturbs travel far, landing about two meters away from him. That's not normal in a terrestrial environment.

Figure 27 - An astronaut hits the ground, and the particles travel freely, falling far away.

In the first part of the video, at minute 17:36, we also see how the astronaut hits the granular soil while walking, and a cloud of dust rises to the height of his knee (see Figure 28). The following sequence shows that moment.

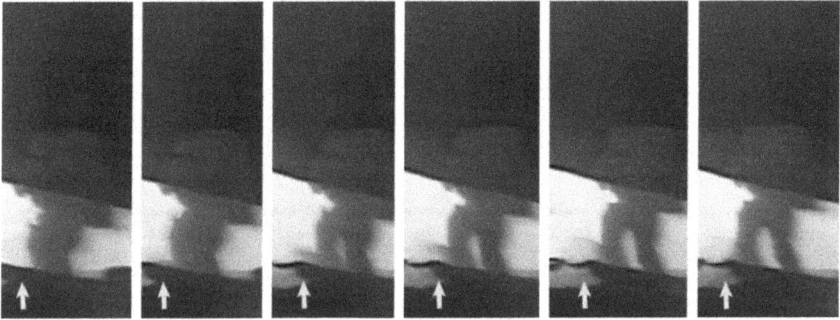

Figure 28 - In the first part of the video, at 17:36, we see the astronaut hitting grains off the ground that rise to his knee; they don't go far.

I invite you to experiment by hitting the sand or gravel with your shoes, for example, on a beach. The large grains go farther, and the small ones, having the resistance of the air, stay behind, forming a cloud, which sometimes lasts a few seconds floating near the boots. For them to reach a distance of two meters, you had to kick them as if you were a professional soccer player. Still, due to a minor stumble, it is not logical for them to reach that far.

What happens when grains are hit in an atmosphere-less, low-gravity environment like the Moon compared to a terrestrial environment? Figure 29 shows a schematic of how soil particles would travel in a terrestrial environment compared to on the Moon.

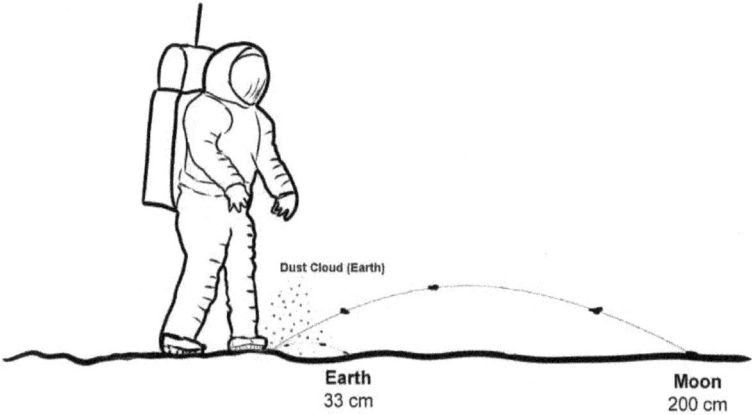

Figure 29 - Trajectory of soil grains when accidentally kicked.

We know that the particles will travel a parabolic trajectory or very close to it. When calculating, if these grains fall 33 centimeters in a terrestrial environment upon being disturbed, then in a lunar habitat, they can reach 2 meters or more. Small grains or dust from the ground in an atmospheric setting, as on Earth, will slow down and form a cloud that does not travel far. On the Moon, small grains fly as far as large grains because no air will slow them down. They will fall farther because there is less gravity on the Moon and no atmosphere.

In the first part of the video, the grains rise, forming a cloud that reaches the astronaut's knees. In the second part, we see with surprise that an accidental blow of the boots shoots grains of soil away from the astronaut. These are two different effects. This indicates that somebody recorded the first part of the video in Earth's environment and the second part on the Moon.

We can draw a similar conclusion to Fact #8:

Fact #9: The NASA video illustrates the different effects of boots hitting the ground. The first part suggests that it was recorded in a terrestrial environment with an atmosphere, and the second part that it was recorded on the Moon.

MOVEMENT AND POSITION OF THE FLAG

We mentioned earlier the experiment that the MythBusters[3] did waving a flag in a big vacuum room. They showed the difference in the pendulum motion of the flag's fabric in air versus a vacuum. Because of inertia, the flag tried to stay in its original position when moving it, so it flapped, as noticed in the NASA videos. We also saw in their experiment that the air contributed to slowing down its oscillation. It is similar to the pendulum movement we observed in the ribbon hanging from the chest of the astronauts' wardrobes. When they saluted the flag, we detected that the ribbon oscillated for a long time. The test performed by the Myth-Busters leads to the conclusion that the moment of raising the flag took place on the Moon, not on Earth. I agree.

However, regarding the flag, there is something curious. If you watch the video in detail, when the astronauts hoist it, the fabric has rigid folds that remain unchanged, even when the flag is moved or pulled from its corners. It waves in a pendulum-like oscillation as if there is no air to stop it quickly, but it has a curve in the lower left corner, a couple of folds in the middle, and an arched shape at the junction with the flagpole. It looks like a starched cloth with predetermined wrinkles to give it that shape on purpose. If NASA wanted to starch the fabric, leaving it plain and not so irregularly shaped would sound more logical. Maybe NASA wanted to give the fabric a particular shape, with folds, that simulate a flag flapping in the wind. Artistically, it looks good, as we see in the photos and the video.

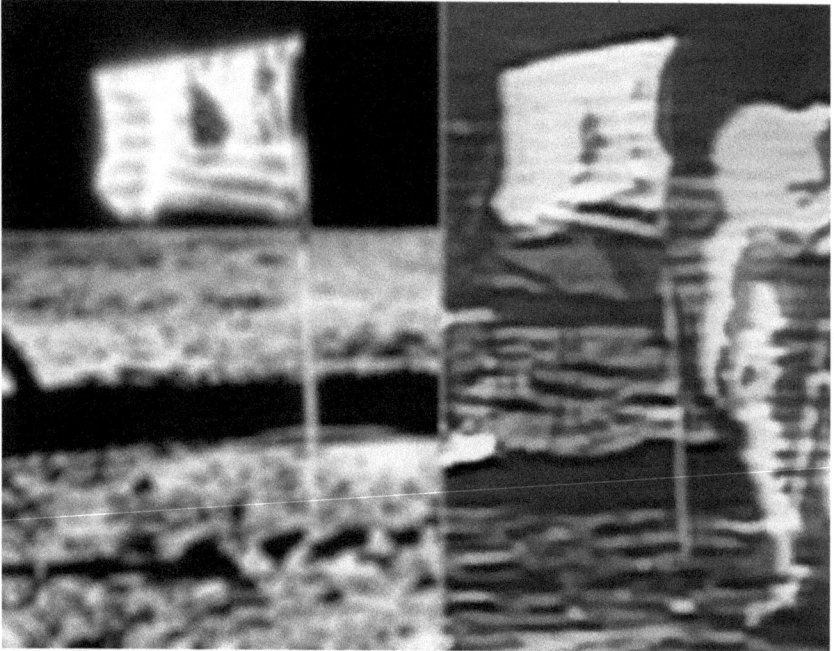

Figure 30 - Comparison of the shape and shadows on the flag. Left is the official photo of the flag, which I converted to black and white and blurred to compare with the NASA video image on the right.

Moreover, the carefully planned shadows and folds in the flag's fabric differ from the video compared to the photos released. Figure 30 compares the official NASA photo of Aldrin's salute to the flag (left), shown in black and white and with a blur, to make it comparable to the image we see from the NASA video (on the right in the figure). You can see that they are similar, but it does not look like the same flag. The astronauts took some time to position the flag with its horizontal support pointing the Sun. It has several folds that produce shadows and illuminated areas. But they do not match. The shots of the flag are from a slightly different direction, as the photo (left) comes from Armstrong's camera, and the one in the video image (right) comes from the TV camera. The astronauts did not move the flag after the salute; both shots capture the moment Armstrong salutes the flag. They should be identical, with the same areas of light and shadow, but they are not. In the video image (right), the flag presents a curved shape at the top and a double arc in contact with the vertical mast. In the NASA photos (left) where we see the flag, only the top arc is visible; the bottom arc is missing. Additionally, it is straight at the top, not curved. There are other details regarding the shape that do not match.

The difference in the horizon's location related to the flag is because the two cameras, the photo and the TV camera, were not at the same height.

In a later chapter, we will analyze the photos of the flag in more detail to determine its authenticity.

About the flag, we can for now conclude the following:

Fact #10: When comparing the photos of the NASA records on Apollo 11, where they show the US flag, with the video on NASA's YouTube channel, we see differences in shape and shadows. They are similar flags, but they are not the same flag.

So far, we can conclude that we have several photos and one video (the second part of the NASA video) probably made by astronauts of an Apollo "Y" mission. They are not from the Apollo 11 mission, and the second part of the NASA video was filmed on the Moon. Whoever filmed the first part of the Apollo 11 video was probably crudely done in a movie studio. For some reason, the video combines an absurd first part with a genuine second part. Why?

In a later chapter, we will assemble all these pieces of this enigmatic puzzle to present one or several hypotheses about what could have happened.

EDITED PHOTOGRAPHS

So far, I have pointed out ten facts, some of which are difficult to explain. I will present more extraordinary findings in the following chapters, and at the end of this book, I will put all the pieces together. In summary, we have:

- The Apollo 11 astronauts could not have taken the Apollo 11 photos. This is a verifiable scientific fact from research[7] I conducted in 2023, which no one has yet refuted with evidence.

- There are 2009 photos from the Lunar Reconnaissance probe showing a lunar module and equipment installed at the Apollo 11 landing site. If the Apollo 11 astronauts did not install it, who did, and when?

- NASA lost the original video stored in magnetic tapes broadcasted in July 1969, when Armstrong and Aldrin supposedly landed on the Moon. We only have a copy shown by NASA on its YouTube channel[15], retrieved from the CBS news archives and other sources. Still, we cannot confirm the authenticity of that copy, as the original video no longer exists for comparison. That video has two parts:

1. The first part is fake; it was not recorded on the Moon, as it shows the horizon tilted, but the astronauts walk vertically. One astronaut kicks the ground as if it were a scene in a terrestrial, not lunar, environment. The astronaut ribbon swung very quickly as if he was on Earth, suggesting the video may have been edited to play at a 60% slower speed.

2. The second part of the video does seem to be taken on the Moon because, at the moment of raising the flag, there is an oscillation movement typical of a fabric being fanned in an environment without an atmosphere. An astronaut kicks the lunar soil, projecting it 2 meters away as it would happen on the Moon. His strap swings several times with a period of oscillation compatible with lunar gravity.

- The flag at the moment of hoisting in the second part of the video is starched, with a predetermined shape, very similar to the one presented in the Apollo 11 archive photos. Still, the shadows indicate that the flag in the official photos and the one in the video are not the same.

- The astronauts carried an external antenna to transmit the TV signal. Still, they never installed it, and the signal sent to Earth was low-resolution. Did they have time to sleep 5 hours but not 20 minutes to install it? The second part of the video shown by NASA on its YouTube channel appears authentic but is of inferior quality. It shows dark interlaced lines which prevent detailed comparisons of the location of the moon rocks and craters with the existing photos.

We know that NASA published photos of Apollo 11 on

the lunar surface a few days after the astronauts' return. These were featured, for instance, in the August 8, 1969 issue of LIFE[19] magazine. Suppose the Apollo 11 astronauts did not set foot on lunar soil. How were those photos obtained, such as those in LIFE magazine and other publications?

In this chapter, I will demonstrate that NASA has manipulated some photos. In the next chapter, I will prove that virtually all of the pictures we find in NASA's Apollo 11 archives are fake and that somebody took them in a movie studio. This assertion may seem like a conspiracy theory, but it is not; it is a conclusion based on facts.

Let's take a look at some pictures.

PHOTOS FROM LIFE MAGAZINE

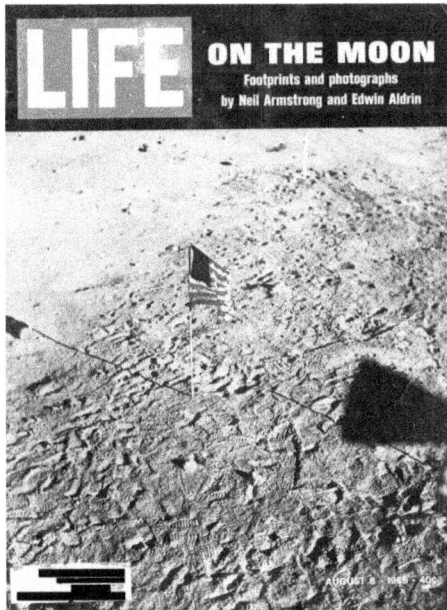

Figure 31 - Cover of LIFE magazine, August 8, 1969.

I have a copy of the August 8, 1969, issue of LIFE[19] magazine, which has an article and several photos about the landing on the Moon by Apollo 11. Apollo 12 and other later missions had yet to happen at that time. So, most likely, the Apollo "Y" Mission would not have happened either. Where were these photos taken? Let's look at the pictures shown in this magazine issue.

There are several photos inside the magazine, many archived by NASA, depicting key moments from the lunar mission. These include Aldrin descending the lunar module ladder, an astronaut installing the seismic equipment and laser reflector system, and the initial salute to the newly erected flag.

The following series of photos appears to be fake.

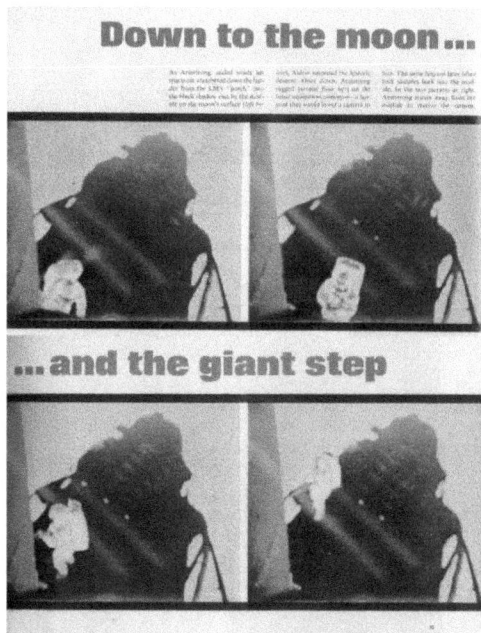

Figure 32 - Sequence of photos on pages 18 and 19 of the magazine. They do not look real.

These are photos taken from above. Are these a few frames from a video taken with the 16mm film recording camera installed in front of the right window that recorded the landing? This camera took this image from a very high angle. Based on the length of the shadows, it looks as if the sun is at about 30 to 45 degrees elevation, not about 15 degrees. In other photos from the spacecraft's windows, such as those in NASA's archives, we see the shadows casting toward the horizon in a more tilted view. Additionally, the silhouette of the lunar module's shadow differs in this magazine compared to the photos on the official site.

A photo edited by NASA appears not only in LIFE magazine but in many publications. NASA admitted to having retouched this photo[20] for publicity purposes.

Figure 33 - Photo AS11-40-5903 retouched by NASA. On the left, as it appears in LIFE magazine. On the right is the original photo. NASA has added a black sky over the astronaut to center it.

I have no problem with NASA retouching photos for publicity purposes, as long as they make it clear in the publications. This photo does not show the antenna the astronauts wear on their suits, protruding from the top of their backpacks. That antenna, which is metallic, flat, and silver, is sometimes visible, depending on whether it reflects the sunlight as if it were a mirror. This effect of the antenna appearing and disappearing has raised suspicions among skeptics.

The astronaut photo and others in the magazine and many publications are the same ones we see in NASA records and the same ones that show the Sun very high, at about 20 to 21 degrees elevation, and not at 14 and 15 degrees. As I will prove, they were made in a film set and designed to show the Sun at 20 degrees elevation, not 14. Perhaps, for unknown reasons, the Apollo 11 astronauts simulated their moonwalk 12 hours early when the Sun was only at a 14-degree elevation.

FAKE FLAG SALUTE PHOTOS

The two photos of Aldrin saluting the flag are faked. Somebody took the first photo in the recording studio. The second is a poorly made copy and fabricated.

Figure 34 - Photo AS11-40-5874. Aldrin raises his right hand
to the level of his face to salute the flag.

The second photo is identical to the first, but Aldrin is in a different position. We can see his face through the helmet since he removed a solar filter that covered it.

If you compare them, the extraordinary thing about these two photos is that all the details are identical. The difference is that the second photo is rotated a couple of degrees. It is reasonable that after a few seconds, Armstrong moved and slightly turned the camera between one shot and the next. However, it would be unusual for him to rotate the camera without also changing its position. How can a chest-mounted camera rotate without shifting a single millimeter to the side or vertically?

Figure 35 - Photo AS11-40-5875. It is identical to photo AS11-40-5874, but Aldrin turns his body and looks at the camera.

In Figure 36, I compared the two photos. Looking at the details near the flag, it is clear that they are identical. The flag pole is one inch in diameter. The distant landscape features do not move to either side by a single millimeter. Who could take a photo, turn the camera slightly, and then, seconds later, take another photo without the camera lens shifting even a millimeter? Wearing a spacesuit and navigating lunar soil would make this precision even more challenging. This finding suggests that the second photo is a montage, based on the first photo. They rotated the image to make it look different, but it did not hide the forgery.

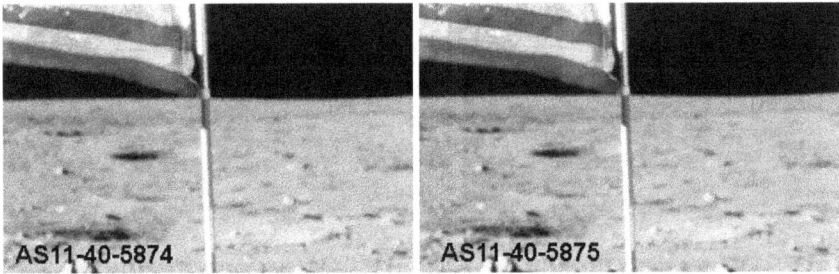

Figure 36 - Comparison of details of photos 5874 and 5875.
They are identical.

We can see another much more conclusive detail to prove that the photo AS11-40-5875 is a composite. Aldrin is facing the Sun. Figure 37 shows the shadows casting from left to right. This means that Aldrin is receiving the Sun's rays directly before him, so the shadows created by the Sun must be congruent with that fact.

Figure 37 - Detail of photo AS11-40-5875.
The shadow on the backpack is very thin.

Having Aldrin rotated to show his face to the camera in photo AS11-40-5874, we see the shadow of his helmet on his spacesuit backpack. The helmet is about the same width as the backpack on their back. Refer to the image of the astronaut with his helmet on, such as the photo in LIFE magazine (Figure 33). So why is there such a thin shadow on his bag?

The shadow of the helmet should be clearly visible on the backpack, occupying a wide area. There shouldn't be a shadow of his hand beyond it if he still has it up. Nor should there be a shadow of his head as if he were not wearing the helmet. The photo is fake; NASA retouched it from the other picture.

Fact #11: NASA has retouched some of the Apollo 11 photos.

Regarding the flag in these photos, we must remember that we have already found that it differs from the one we see in the video. The shadows and shapes are dissimilar.

The photos show the Sun very high, and it does not seem that a photographer took them on the Moon. Some appear to be illustrative, but not from the Moon. There are at least a couple of photos that NASA retouched. These photos, which are of outstanding quality, show craters and many details that make them look very realistic as if they were taken on the Moon. But were they shot on the Moon's surface? The following chapter will prove they were produced in a cinematographic studio. Again, this is not just another conspiracy theory but rather a study using factual evidence.

IN A FILM STUDIO

In this chapter, I demonstrate that a photographer took most of the Apollo 11 photos in a ground-based movie studio. The astronauts did not take them on the Moon. How did I conclude this? Several clues lead me to this fact.

Clue #1: Different Craters.

The craters depicted in the official Apollo 11 photos are clearly different from those visible today in the Sea of Tranquility. Recent images captured by the Lunar Reconnaissance probe show the craters around the base of the Apollo 11 Lunar Module at the landing site. They are similar yet distinctly different craters.

Clue #2: The lunar module is in a different position.

The base of the lunar module on the Moon, which the Lunar Reconnaissance probe photographed in 2009, is in a different position from the lunar module we see in the official NASA photos of Apollo 11. The spacecraft on the Moon is offset 4 to 6 meters to the east. Somebody made a very good replica of the Apollo 11 landing site. Still, the Apollo "Y" did not land exactly in the correct location.

Clue #3: We see a fake horizon wall in the landscape.

Stereoscopic images show us that in the landing place we see in the Apollo 11 photos, there is a backdrop simulating the horizon.

Clue #4: What illuminates the landscape is not the Sun.

From the analysis of the sharpness of the shadows, we conclude that it is a cinematographic lamp that illuminates the landscape and produces different penumbra sizes depending on the distance the astronauts are from it.

CLUE #1: DIFFERENT CRATERS

It is very suspicious to find in the official Apollo 11 photos that there is a missing crater. That crater was only 10 meters from the lunar module, which, on landing, may well have fallen there. It's called the Double Crater, but it looks like, for some reason, it is not double but single in the Apollo 11 photos.

We had already mentioned that the Lunar Reconnaissance probe took pictures of the Apollo 11 landing site (Figure 7), showing the lunar module, some of the instruments installed there, and the footprints left by the astronauts at the site. Some of the photos were from the crater called "Little West Crater"[21]. Figure 38 displays a panoramic view created from a mosaic of photos taken by Armstrong, who walked 60 meters from the lunar module to capture these images. This wide-angle composition was achieved by superimposing several photographs.

Figure 38 - The Panoramic view from the "Little West" crater is 60 meters east of the spacecraft. The astronaut walked there and took pictures from that place.

However, for reasons that are not clear, we do not see photographic records in 1969 of the astronauts walking near the "Double Crater," which was only 10 meters from the spacecraft. If they walked 60 meters to the Little West Crater, why did they not show the Double Crater a few steps away? It's a double crater at least one meter deep and 15 meters in diameter. It is named "double," but it is triple because it has a smaller third crater in the middle. When Neil Armstrong took photos of Aldrin descending the Lunar Module's ladder, he was close to the crater's edge. We found some images released recently of these craters. Still, for reasons unknown, there was no photographic evidence in 1969 of these craters very close to the spacecraft. NASA ignored these craters even though the spaceship could have landed inside them and probably put the mission in danger. We should have photographs of that crater, perhaps taken if an astronaut had walked to the far edge to capture a good shot, showing the Double Crater in the foreground and the Lunar Module behind it. But this was not the case. Yes, astronauts took pictures from the Lunar Module cockpit and some from the surface where we see it. Photos were released recently, but the crater is not double. One half is 1 meter deep, but the other is almost non-existent. It does not match the recent NASA photos taken by the Lunar Reconnaissance probe.

The following figure shows another photo from the Lunar Reconnaissance mission[6] that NASA published. In this photo, the sun is very low on the horizon, to the west, and we can appreciate the shadows well, which allows us to estimate the depth of the craters.

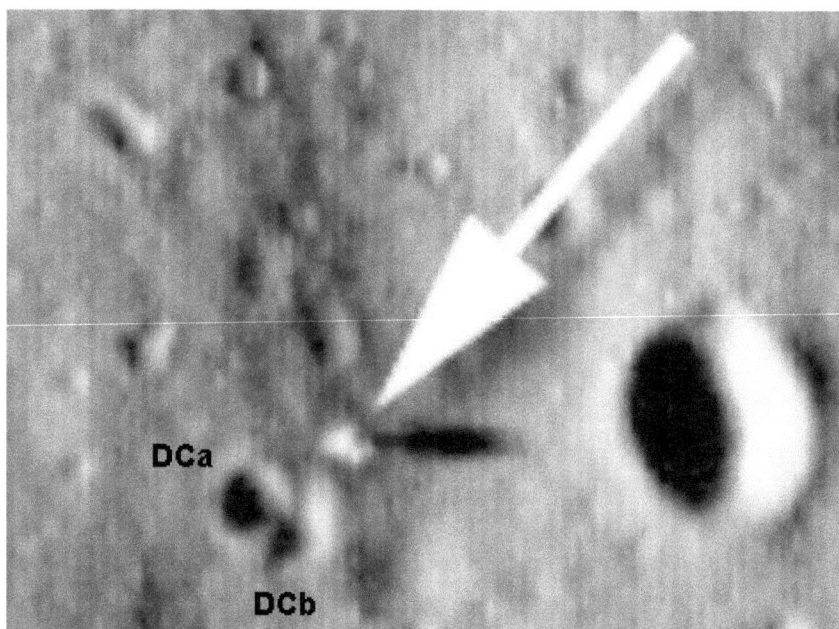

Figure 39 - Lunar Reconnaissance probe photo of the Apollo 11 landing site. Double Crater, DCa, and DCb are near the lunar module. The arrow points to the base of the lunar module. The crater to the right in the photo is "Little West."

The base of the Lunar Module, the section that remained there after the astronauts returned, is about 3 meters (10 feet) high. In the photo, we see the shadow it casts. Comparing the lengths of the shadows of the Double Crater, they are about one-third the length of the lunar module shadow. It indicates that the crater is about 1 meter deep in DCa; in DCb, it is at least 80 centimeters deep and possibly as much as 1 meter deep too.

In Figure 40, we see another shot showing the same crater again. The sunlight comes from the other side, from the east, as it did during the Apollo 11 mission. We see both craters (DCa and DCb), which form the Double Crater and are of similar depths. And we see a third minor, shallow hole in between them. This small crater and the two craters of the double system form a trio.

Figure 40 - Another Lunar Reconnaissance probe photo of the Apollo 11 landing site. Now, the Sun is in the east. The Double Crater near the Lunar Module is two depressions of similar depth.

We do not see photos of the astronauts walking near the Double Crater. Still, we see pictures taken by Armstrong and Aldrin from the spacecraft's windows when they landed. In four of them, taken in black and white, we can see the Double Crater next to them: photos 5741, 5742, 5743 and 5744. But, surprisingly, half of the Double Crater looks plain. We see the DCa portion, but the DCb side does not exist or is just a depression about 10 centimeters deep. We do not see the third crater between them; it is just a smaller crater on the rim and in another position. The images of this double crater in the Lunar Reconnaissance photos do not match those we see in the pictures from the Lunar Module window. Why? The crater DCb is virtually nonexistent, as is the small crater in between them. They are two different sites.

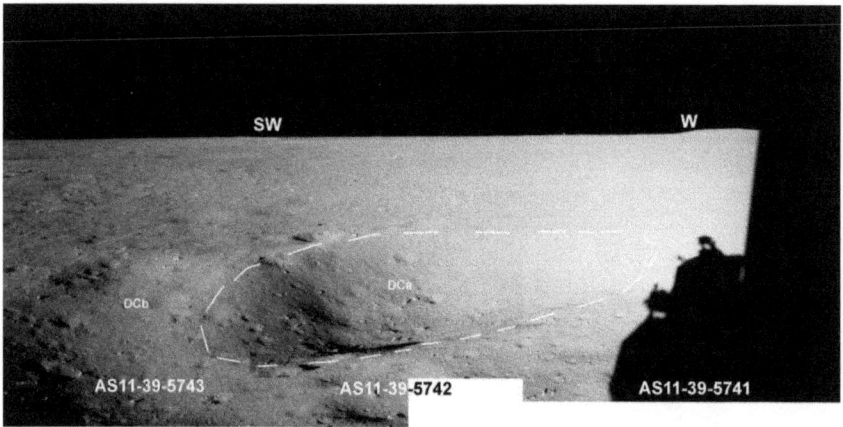

Figure 41 - Mosaic of photos 5741, 5742 and 5743.
The double crater looks simple.

Also intriguing is the presence of a square crater that should not exist (Figure 42).

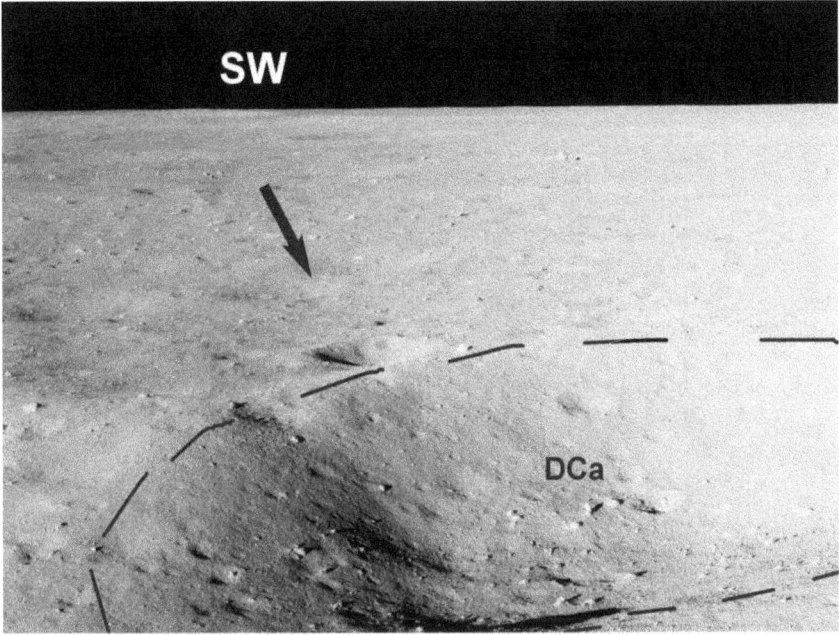

Figure 42 – Presence of a mysterious square crater. The top image is from official Apollo 11 photos. The bottom is a photo from the Lunar Reconnaissance probe. There is a significant difference.

There is a crater in the shape of a square at the edge of crater DCa. See the arrow in Figure 42. Craters are typically round, not square. This crater is an exception, but the issue is that it does not exist on the Moon according to NASA photos from the Lunar Reconnaissance probe. The square crater is visible in the official Apollo 11 photos (Top photo in the figure), taken towards the southwest in black and white film from the cabin when the Lunar Module landed. This strange crater is located beyond DCa, one of the craters in the "double crater" system. In the Lunar Reconnaissance image (Bottom photo in the figure), we find two smaller round craters not at the edge of DCa. These photos are not from the same place.

On the Moon, there is no atmosphere, running water, or wind. No erosive forces will change the Moon's surface to such an extent in just 40 years. The only logical explanation for the big differences in the crater configuration around the base of the Lunar Module is that we are looking at a fabricated landscape in the Apollo 11 photos. It must be a replica made in a desert or underground studio, where the person in charge of digging the crater could have done a better job or needed reliable information in 1969 about the Moon's surface configuration at the Sea of Tranquility.

These findings lead us to the following conclusion:

Fact #12: In the NASA Apollo 11 pictures, we see significant differences in the craters around the base of the Lunar Module towards the southwest, comparing them against images from the Lunar Reconnaissance probe.

CLUE #2: THE LUNAR MODULE IS IN A DIFFERENT POSITION.

Looking at the 1969 photos of Apollo 11 in NASA records, the location of the Lunar Module we see in these photos does not match what it is currently on the Moon. Somebody made a simulation in a place on Earth replicating the Apollo 11 landing site, and they installed a lunar module. Still, the location of that spacecraft on that fake site differed from the one we see in the Lunar Reconnaissance probe photos.

Suppose someone created an area similar to the Apollo 11 landing site in a cinematographic studio, underground setting, or desert. In that case, we cannot expect it to be perfect. As mentioned in the previous section, they could have recreated the craters more accurately. And suppose there was an Apollo "Y" mission that later installed all the Apollo 11 equipment. In that case, we cannot expect it to land precisely where they projected it to touch down.

How do you create a film studio identical to the moonscape at the Apollo 11 landing site? Apollo 10 flew as low as 14 kilometers[22], taking pictures with the Hasselblad 500EL camera, adapted for NASA, which engineers also use for photogrammetry on Earth's surface. Apollo 10 took pairs of photos that allowed stereoscopic vision to determine ground heights. In comparison, the Lunar Reconnaissance probe flew at an altitude of 50 kilometers. So, 14 kilometers gave NASA good terrain details to plan the Apollo 11 landing.

With these photos, NASA was able to create an accurate landing map that they could use to make an excellent replica of the lunar surface at the selected site. This 1:1 scale replica would be ideal for simulations, allowing astronauts to familiarize themselves with their landing area and equipment installation sites in advance. The Apollo 11 photos are from that replica. The issue with using the Apollo 10 images to create this replica is that the sunlight only comes from one direction, from the east. Apollo 10 could not stay there for a long time, just a few days, and then returned to Earth. It is challenging to accurately determine the dimensions and shapes of craters and other lunar details when they are illuminated from only one direction. The Lunar Reconnaissance Orbiter remains in orbit for a long time, allowing us to observe the lunar landscapes under various types of illumination, giving good details of the Moon's surface.

I conducted experiment[23] by creating a clay replica of the landing site. I used the Moon maps on Google Earth, which are based on the images taken by the Lunar Reconnaissance Orbiter.

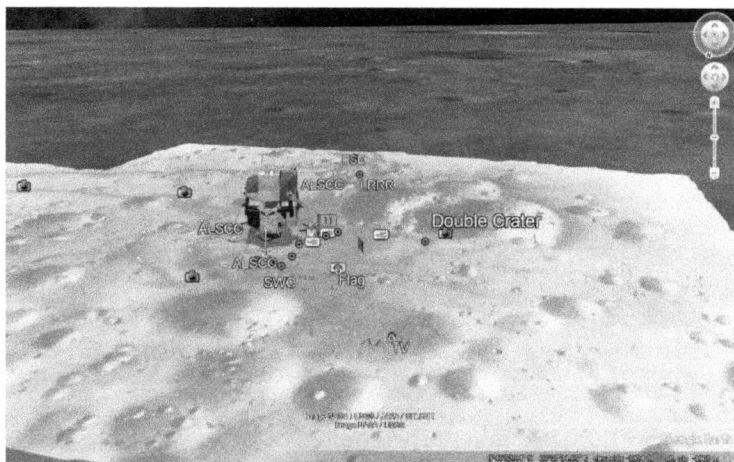

Figure 43 - Image from the Google Earth application showing the Apollo 11 landing site. We see the Double Crater on the right. The view is to the south.

Let's study the landing site in detail. The Google Earth application shows the lunar surface, the landing place of the Apollo 11 mission, and a good representation of the Lunar Module in three dimensions. The application credits the images provided by NASA and the USGS.

Google Earth uses images from the Lunar Reconnaissance probe. So, we see a good representation of what exists on the Moon. Let's see if it matches what we find in the pictures.

Using the available Lunar Reconnaissance photos and Google Earth images, I have created the following map of the current location of equipment at the Apollo 11 landing site as it is today on the Moon (Figure 44). NASA could not locate the flag in 2009 photos. It appears to have been lost after the liftoff of the top section of the Lunar Module when returning to rendezvous with the orbiting service module. I estimate its original location on this map.

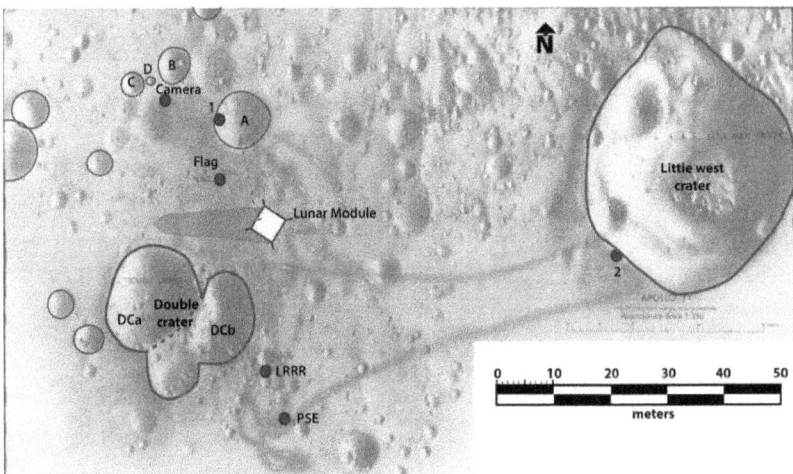

Figure 44 - Apollo 11 landing site map as it currently stands on the Moon.

I drew the camera's location and some installed equipment (LRRR and PSE) on the map. The point marked "2" is where the astronaut took several photos to make the photo mosaic from Little West Crater. At "1" is the location where Armstrong should have been standing when he took a picture of Aldrin saluting the flag. Armstrong must have been precisely to the north, as the Sun casts shadows to the west, and Aldrin faces the Sun. Did Armstrong stand inside crater "A"?

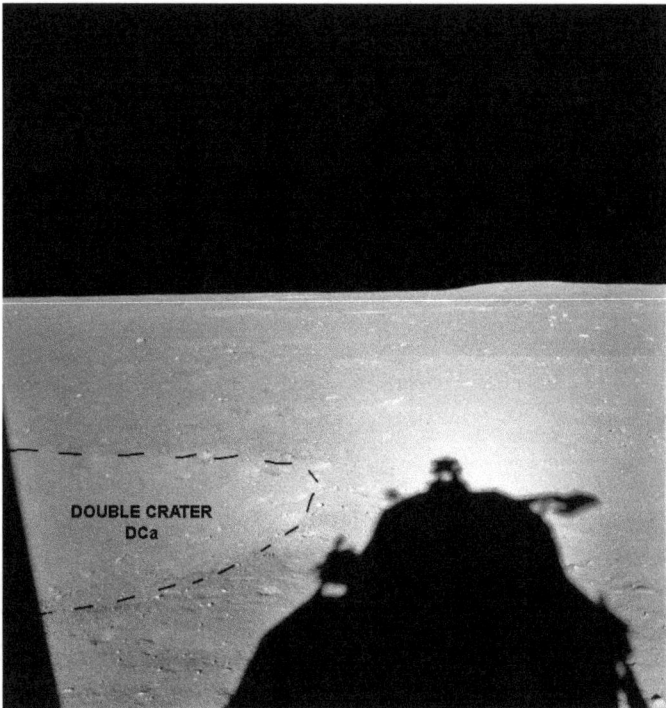

Figure 45 - Photo AS11-37-5454 cropped and contrast-enhanced to see the rim of crater DCa.

The entire Lunar Module is about 6 meters tall (20 feet). After analyzing 12 photos, I found that the average elevation angle of the Sun was 20 degrees. Thus, the shadow should be 16.5 meters long. Then, the shadow should reach to coincide with the center of crater DCa, which is the west side of the Double Crater. However, in the Apollo 11 photos, it goes further, almost to the western edge of the crater (see Figure 45). Why doesn't it coincide? The answer is that the lunar module at the hypothetical training site was located about 4 to 6 meters (20 feet) further west. If so, what we see on the Moon does not match what we see in the photos.

Someone might think that the Sun could have been lower, and the shadow would have been longer, reaching the end of the Double Crater. However, the study of 12 photos showed that the Sun, as we see it in these photos, was at 20 to 21 degrees elevation.

Further proof that the spacecraft is displaced is more compelling. It is provided by looking again at the mosaic of photos taken from the Lunar Module window and comparing it with the Lunar Reconnaissance photos (Figure 46).

Figure 46 indicates the southwest (SW) direction. We can find the west position by looking at the brightness of the Sun in Figure 45. The Sun casts its shadows just to the west. The Moon's rotation axis is not tilted like the Earth's, and the Sun will always point to the west or east. There are no seasons on the Moon. The astronaut took the pictures on the Moon with two Hasselblad cameras, one on the capsule with a lens of 80mm and without the Réseau plate, and the other one during the extravehicular activity, with a lens of 60mm and holding the Réseau plate[11]. Apollo 11 carried four cameras[25], two of them in the Lunar Module. We know the photos in black and white taken from the cabin were shot with a Hasselblad camera using an 80mm lens that produces an angle of view[24] of 38°.

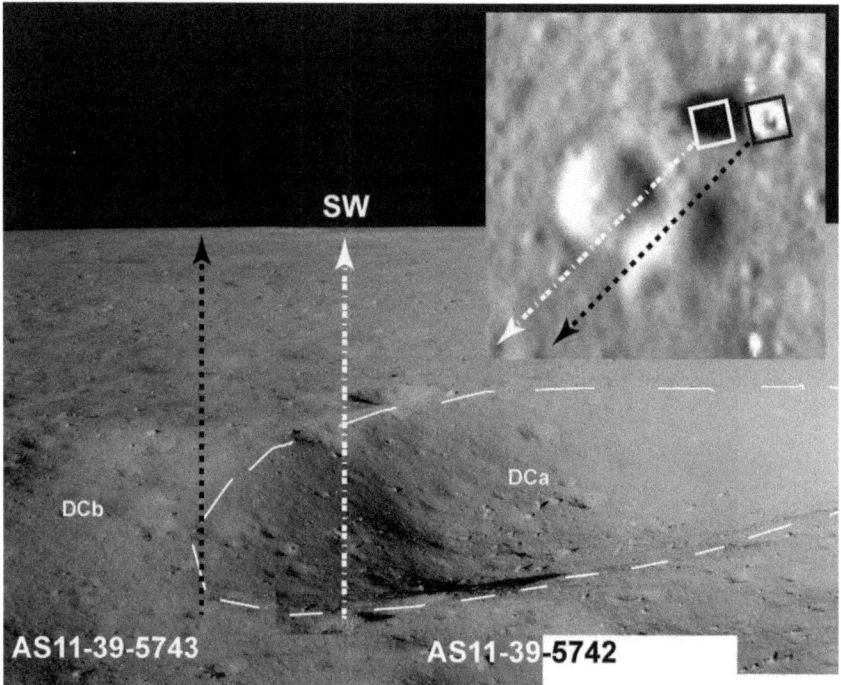

Figure 46 - Mosaic of photos 5742 and 5743 compared to Lunar Reconnaissance photo (inset). The SW direction passes over crater DCa but should
pass over the left rim. The spacecraft is offset.

Using the camera's field of view and the location to the
west, I was able to measure a 45-degree angle to pinpoint
the exact location of the southwest (SW).

The photo mosaic showed that the SW direction passes
over crater DCa. But in the Lunar Reconnaissance photos,
it should pass over one of its ends at the junction between
DCa and DCb (see Figure 46). So, the Lunar Module that
landed on the Moon, as seen in the actual photos of the
lunar surface, is about 4 to 6 meters to the east, not where
the official Apollo 11 photos indicated.

Let us now analyze the photo AS11-37-5481, Figure 47.
We see craters B, C, and D (see them on the site map in

Figure 43). The stand with the TV camera is in the direction of the small crater D. That matches what we see on the map. However, we do not see crater A in the Apollo 11 photo. It is not a tiny crater and should be visible in the image.

Figure 47 - Analysis of AS11-37-5481 photo and nearby craters.

The photo AS11-37-5481 from the capsule window must show crater A. We can confirm it using Google Earth images in Figure 48. The window is the small triangle at the upper right of the lunar module. This drawing demonstrates that crater A is in the line of vision from the capsule window. The small nozzles of the spacecraft are in the same line of sight. Is this another crater that someone forgot to dig in the ground of the moonscape in the recording studio? It's possible. Or the spacecraft was located in that film set about 4 to 6 meters to the west, as we had already deduced by studying the shadow of the Lunar Module. Consequently, that crater could not be observed from the window.

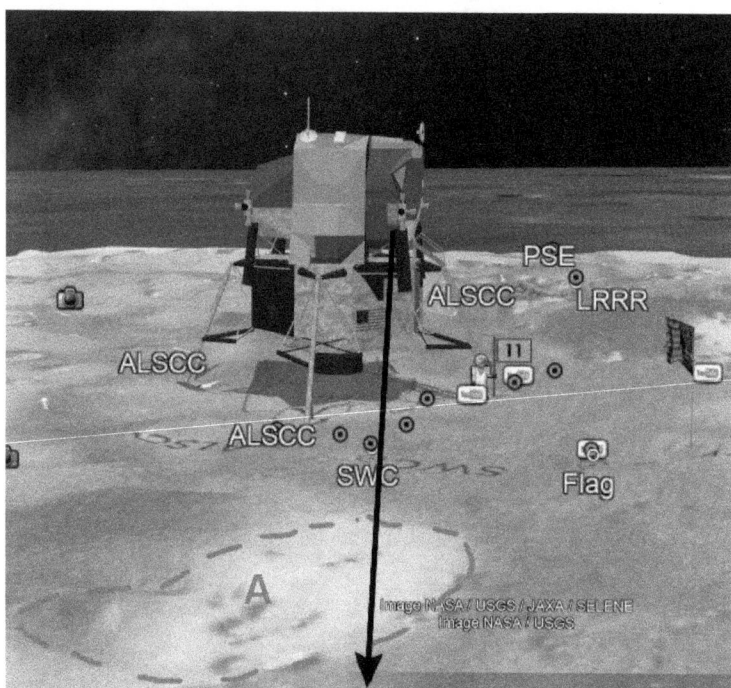

Figure 48 - Google Earth image. Crater "A" must be visible from the Lunar Module window.

In Figure 49, I present the map of the cinematographic studio model. This map shows where a photographer took photos of Apollo 11 in an Earth-based studio, not on the Moon. The Lunar Module is displaced 4 to 6 meters to the west. Therefore, the position of the flag and the place where Armstrong stood to take the picture of Aldrin saluting it has also shifted. Now we see Armstrong standing on solid ground, not inside crater A. I have not drawn crater DCb in the film studio map because it does not exist.

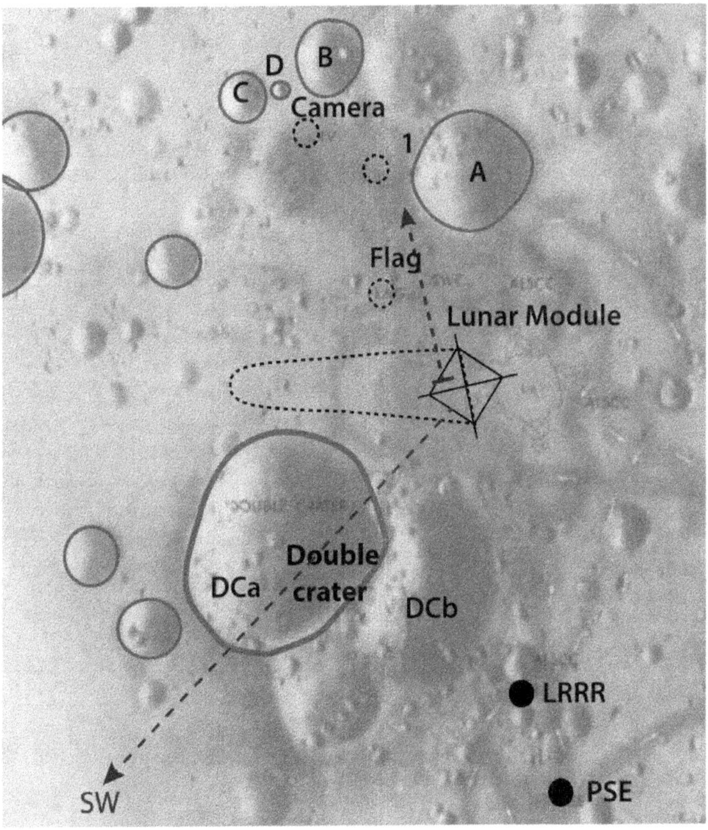

Figure 49 - Map of the artificial terrain created in the recording studio.

In conclusion, the photos of Apollo 11 do not match what is seen on the Moon today. The location of the Lunar Module has shifted around 4 to 6 meters.

Fact #13: The location of the Lunar Module varies by 4 to 6 meters. Its position on the Moon's surface today differs from what the Apollo 11 photos show.

Maybe someone with good film experience advised NASA to create a big studio like the ones in Hollywood movies. There, they took videos and pictures. They did it possibly in a large underground cavern and away from the curious witnesses. Or they built it in a remote desert area, taking only night shots. Based on the photogrammetric photos obtained by Apollo 10, they created the entire environment, digging craters to simulate the lunar landscape. In doing so, they underestimated the depth of the DCb crater, resulting in a formation that did not resemble the actual Double Crater on the Moon, which might be more aptly called a triple crater. If we look at the Moon terrain in this location with the sunlight coming from the east only, we may assume DCb is not a crater but a little hill that casts shadows towards the west. But if we look at the same area, with the sunlight coming from the west, we may notice there are two craters, not one. The creators of the cinematographic studio might be confused as they only have photos showing the terrain lit by sunlight from the east, not the west.

The map I made from Lunar Reconnaissance photos shows the Apollo "Y" landing spot on the Moon, not Apollo 11, which was never there. Apollo "Y" landed 4 to 6 meters away (13 to 20 feet) to the east of the target.

Someone arranged all the equipment in the cinematographic studio and recorded videos of astronauts walking. Today, we see a mixture of a fake movie (from the studio) in the first part of the NASA YouTube video, which features a cropped horizon, and an authentic one in the second part (Apollo "Y") despite its poor resolution.

Also, they took photos in the film studio set. Were all the pictures we see today taken in this studio? Or are some of them mixed in from the mysterious Apollo "Y" mission? Most of them are from the recording studio. If a mix of photos shows details of rocks and little craters, the difference would be prominent when looking at the studio photos versus the images from the mysterious Apollo mission. I have found differences in the Double Crater, but there are many more differences than anyone could detect. One photo in the Apollo 11 collection captures a view towards the dark sky where we see the beautiful Earth behind the top section of the Lunar Module. This image might be from Apollo "Y" since it does not show the ground.

The designers of the cinematographic studio used Apollo 10 photos, which may have good details of craters but could have been more helpful for smaller pieces, like rocks. Future Moon missions, with a higher level of detail, will again prove that the moon landscape at the Apollo 11 landing site does not match what we see in the Apollo 11 photos.

CLUE #3: WE SEE A FAKE HORIZON WALL IN THE LANDSCAPE.

There is a high probability that somebody took the Apollo 11 photos in a film set. If this is a simulated site, how big is it? Are there any other details that tell us this is so? Clues 3 and 4 give us more information.

Simulating the landing site may require faking the horizon, too. How can the horizon be manufactured? It is feasible to make craters matching those existing on the Moon, based on the photos taken by Apollo 10 before the Apollo 11 mission, but how do you simulate the profile of distant mountains or hide the existing horizon at the recording site if it is, for example, in a desert? The way to do it would be to put up a fake wall where a painter draws the distant profile of the landing site, as it must be in the Sea of Tranquility on the Moon. It could be a wooden or fabric curtain.

Some photos, especially those taken from the windows inside the lunar module, elevated about 5 meters, show a discontinuity in the horizon as if there were indeed a stage. We see a detail of it in Figure 50.

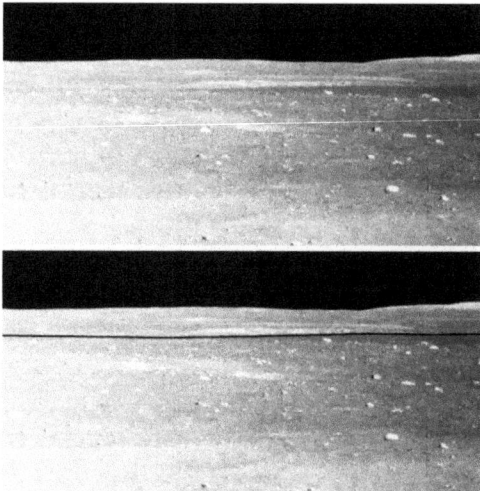

Figure 50 - Possible backdrop with a drawing of the lunar horizon.

In the figure, the upper part shows a fragment of one of the photos taken from the Lunar Module cabin. The lower image includes a line that I think separates the natural environment with craters from the environment illustrated on a wall in the background. Near the horizon, there is a discontinuity that resembles mountains or a distant lunar horizon, as if painted on a vertical wall by an artist.

Is it a coincidence? Is there a section of the lunar floor that descends abruptly and forms that discontinuity? Or is it a screen set up there, perhaps in the desert, about 200 meters from the camera?

To verify this, we could observe the terrain in three dimensions. If we could see the depth of each detail, identifying which stones are closer and which are farther away, we could determine whether the backdrop is a fixed distance from the camera or a lunar landscape that stretches indefinitely toward the horizon.

We observe in three dimensions through our stereoscopic view because we have two eyes. Can we see the three dimensions on the Moon's surface? Yes, we can, using the two eyes of the spider!

Let's see the spider:

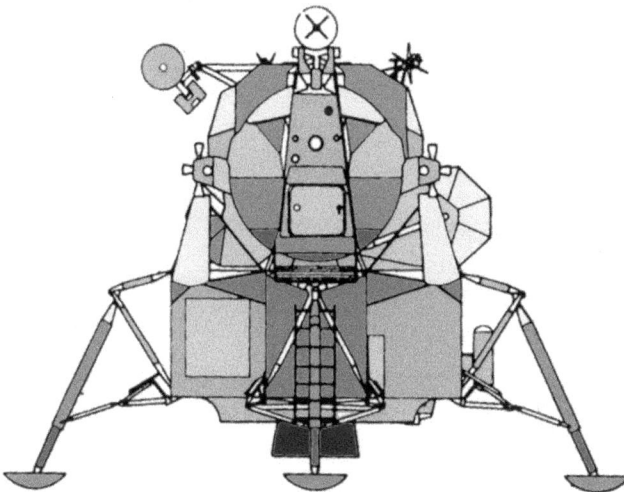

Figure 51 - The Lunar Module, which looks like a spider, has two windows at the top (in the shape of triangles).

The Lunar Module has three windows, one on the roof and two at the front. The two front windows, in the shape of triangles, look like the eyes of a spider. Using this metaphor, we can achieve stereoscopic vision by using either a pair of eyes or photos taken from the Lunar Module, one from the right window and one from the left window (from each eye of the spider), separated by 1.3 meters (4.3 feet).

I used photos AS-39-5741 (from the left window) and AS-37-5489 (from the right window). Both images pointed toward the west.

With them, I prepared the stereoscopic pair shown in Figure 52.

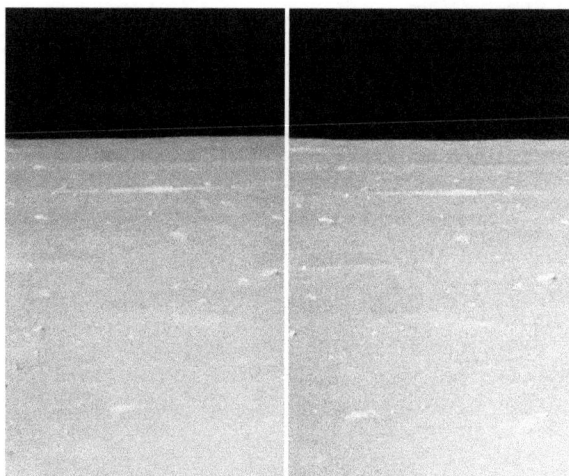

Figure 52 - Stereoscopic pair using the two photos
from the Lunar Module windows.

Some people can see three dimensions with the naked eye when looking at this pair of images. The technique is to focus with the left eye on the left image and the right eye on the right image. Another way to do this is to use a stereoscope, similar to the one assembled in experiment #10 in our book "Researching a Real UFO"[26], where we used a pair of magnifying glasses on its construction. I wrote this book with Chris Lock a few years ago.

I posted these images in a Facebook group dealing with stereoscopic images. The reactions were different, and some were very emotional. A few felt that, in reality, the Apollo program was a big lie. However, a majority immediately rejected any suggestion that we were looking at a backdrop created in a film set, citing photos from the Lunar Reconnaissance Orbiter photos as definitive evidence that Apollo 11 was authentic. Some took a more neutral stance, stating that the conclusion is not evident and either interpretation is valid.

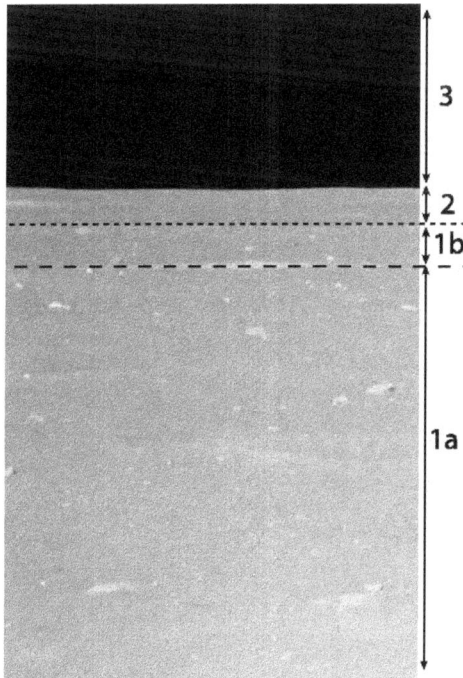

Figure 53 - Different areas can be seen in the photos when viewed in three dimensions.

Looking at this pair of 3-D photos, I see what I describe in Figure 53 (even though some still insist that the Apollo program was authentic). Zone 1a indicates a close terrain full of rocks extending backward. Zone 1b continues zone 1a, but the terrain slopes down and rises again. Zone 2 is flat and does not gradually extend into the distance. It resembles a flat surface (curtain) where an artist drew a lunar terrain simulating the lunar horizon. Zone 3 is an entirely black sky.

Is it a distant wall? The interpretation of these images in three dimensions varies depending on the viewer. In that case, there is a method that can provide us with an objective conclusion about what we are observing: the use of the parallax. Parallax measures the distance between points and noting the difference, which I illustrate in the image below.

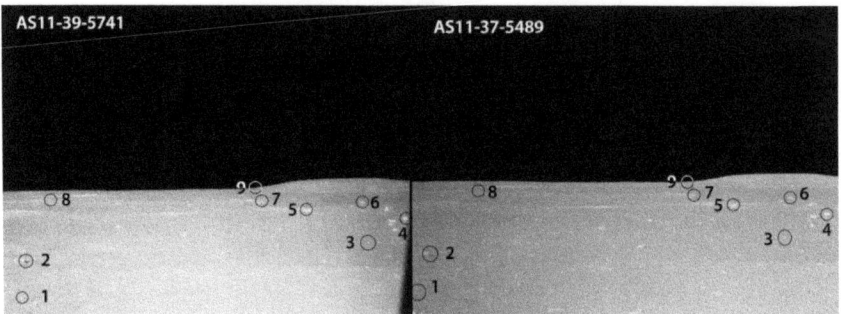

Figure 54 - Key points within the two analyzed photos to measure parallax.

We found different rocks or details in each photo to use for this analysis. I have numbered them from 1 to 9. These dots represent details of the terrain. For example, point 2 shows the right edge of a rock.

I enlarged these images on my computer screen and

measured the separation between each pair of points (parallax) in millimeters. I present the results in Table 4:

Point	Separation
1	231
2	236
3	242
4	244
5	247
6	249
7	251
8	251
9	251

Table 4 - Separation in millimeters between each pair of points.

In the values of the table, we observe how the separation between each pair of points increases as the point moves away. Points that are near are closer together, and distant points are farther apart. The last 3 points, 7, 8, and 9, clearly do not follow the sequence but seem to be at the same separation. That means those three points are at the same distance from the lunar module. They are not details on a distant terrain, but somebody drew them on a backdrop.

Some people may still doubt the results despite this evidence. However, many clues and evidence can enlighten us about what happened with Apollo 11. This book presents substantial evidence that may well be reviewed or debated.

Fact #14: Stereoscopic analysis of Apollo 11 photos indicates the presence of a distant screen that simulates the lunar landscape.

CLUE #4: WHAT IS ILLUMINATING THE LANDSCAPE IS NOT THE SUN.

If this is a cinematographic studio, how did someone simulate the sunlight? The answer is a sizeable cinematic lamp. But that lamp would have to meet certain conditions. It must be adjustable in height, so the people involved in constructing this site installed the studio lamp on a mechanism that could raise and lower it. They chose 20 degrees of elevation for the photographic sessions. Additionally, it met another essential requirement: the illuminated area had to be large enough to mimic the effects of the Sun's shadows.

The characteristics of shadow and penumbra must meet specific criteria. Only some realize the effect of shadows created by the Sun. The Sun has an apparent size of half a degree in diameter in the sky. That is the same value on Earth and the Moon since they are approximately the same distance from the Sun. The shadows are blurred at the edges because it is not just a spot but a luminous disk. Additionally, the shadows become blurrier as the distance between the object and the shadow it casts increases. Suppose you look at a tall pole and its shadow on the ground. In that case, you will notice that the shadow seems well-defined near the base of the pole. At the same time, as you move away from the base, you see the shadow getting blurrier and blurrier, with edges that are less defined.

When you look at the photos of the helicopter NASA

sent to Mars, you notice the shadow this flying device casts on the ground. The shadows are very sharp. That is because the Sun on Mars looks small in the sky and does not cause the same level of blurriness as on Earth, which is closer to the Sun.

How do you simulate the size of the Sun in a studio to create the same shadow effects that you would have on Earth or the Moon? It's simple: the size of the lamp, as seen from the Lunar Module, must be calculated to make the same effect. For example, suppose the cinematographers place the light at 140 meters (460 feet) from the Lunar Module replica. In that case, this lamp must measure 1.2 meters (3 feet) in diameter. You can calculate this size with the tangent of half a degree multiplied by 140 meters. The light can be a set of several lamps joined together to form a circular surface of 1.2 meters in diameter.

But what happens if the objects casting shadows are at varying distances from the lamp that simulates the Sun? Well, the shadows change in their penumbra value. That is what happens in the Apollo 11 photos, and it leads us to conclude that the light is not from the Sun but from a large film studio lamp.

The following figure describes the phenomenon of shadows due to the Sun.

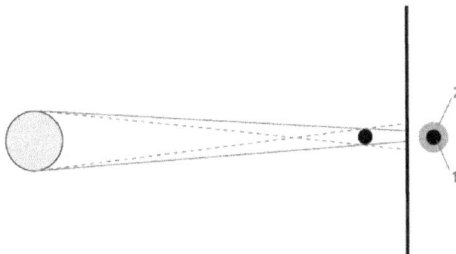

Figure 55 - Effect of the shadows cast by the Sun. Zone 1 is the shadow, and zone 2 is the penumbra.

The light source is on the left. The straight line represents the ground. An object, for example, the astronaut's helmet, is between the ground and the light source. The shadow zone on the floor, labeled as number 1, is an area where no light reaches. That is why it looks black. Marked with 2 is the area where light partially reaches, and that's why it appears gray. And outside these zones, everything is illuminated. The "penumbra," zone 2, is blurry and transitions from a dark part in the central zone to a lighter piece in the extreme.

The penumbra zone increases in size if the object gets closer to the light source and decreases if it gets farther away. When the light source is the Sun, no matter how we move, for example, walking in a large square, we will not be closer or farther away from the Sun. The shadow of our head cast on the floor will always be equally blurred. But if the light source is a nearby artificial light and not the distant Sun, will moving closer or farther from the lamp alter the size of the penumbra zone? Well, that is happening in the Apollo 11 photos, indicating what illuminates the landscape is not the Sun.

Figure 56 - Location of the astronaut when he took pictures where his shadow can be seen on the ground.

There are three places where the astronaut took pictures, and we can see the shadow of his body cast on the ground. The photos are AS11-40-5961, taken when the astronaut walked 60 meters east to the edge of the Little West crater; AS11-40-5962, captured while the astronaut is at an intermediate point between that crater and the Lunar Module; and AS-40-5882, taken as the astronaut walked westward, almost to where the Lunar Module's shadow ends on the ground.

I show the three photos in Figure 57.

Figure 57 - From left to right, photos AS11-40-5961, AS11-40-5962, and AS-40-5882.

We find the astronaut's shadow on the ground in these photos. And in them, the Sun, or movie studio lamp, is at 20.3, 21.6, and 20.6 degrees of elevation, respectively, according to my findings. The difference is due to the irregular horizon I used to measure the angles. Still, they all show the Sun is very high, at 20.8 degrees on average (NASA says that during the Apollo 11 landing, these angles should be between 14 to 15 degrees). If a lamp was used, it could go up or down to compensate for the astronaut's proximity to it. However, the lamp cannot move closer or farther away from the astronaut to compensate for the effect of the penumbra. Why this conclusion? Because the penumbra zone is different in these three photos.

Figure 58 shows the details of the penumbra zones. I processed the images, turning them black and white, cropping them, and making a Shadow/Highlights effect in Photoshop to increase the contrast between the shadow and the penumbra zone. I used the same Photoshop parameters in all the photos, and anyone can replicate this experiment.

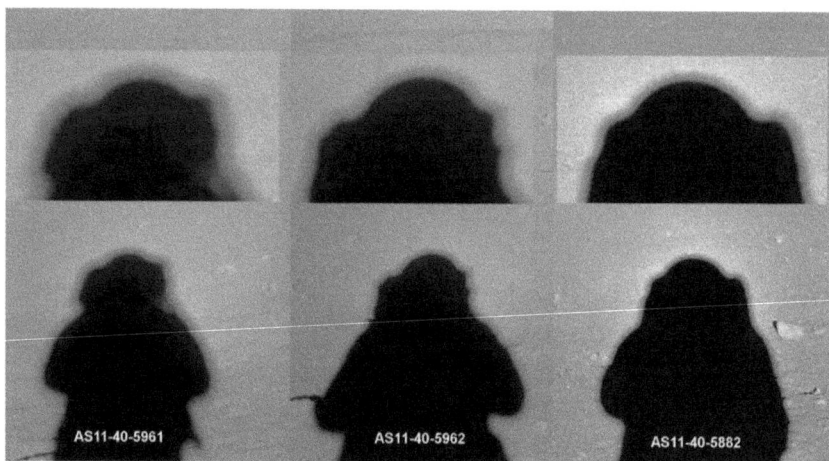

Figure 58 - Shadow and penumbra on each photo.
The closer to the lamp, the larger the penumbra zone.

In the images of the astronaut's shadows, it is clear that as he moves away to the west, the zone of penumbra, which we see here as a gray halo around the shadow of his helmet, becomes narrower. The distance on the ground between the first photo, labeled 5961, and the last, labeled 5882, shown in this analysis, is about 80 meters. Notably, the penumbra zone of the first photo is twice as wide as that of the last image. That gives us an estimate that the cinematic reflector was about 80 meters from the rim of the Little West crater. So, the lamp was installed nearly 140 meters from the Lunar Module in the recording studio. And that reflector could go up and down. Still, it couldn't move horizontally, or they didn't consider it essential.

It is important to note that in the first photo, labeled 5961, the astronaut appears on slightly elevated ground. Therefore, his shadow is a little farther away from his head. That causes the effect of a larger penumbra to occur. However, the difference in that distance is at most 10% and, therefore, would not explain why the penumbra is twice that of photo 5882.

If the astronaut is illuminated by the Sun rather than a cinematographic reflector, the penumbra zones would be the same due to the Sun's distance. There would not be such a significant variation.

This clue leads to the conclusion:

Fact #15: Analysis of the shadows of the Apollo 11 photos, in their penumbra dimension, shows us that what illuminates the landscape and the astronauts is a cinematographic reflector, perhaps 1.2 meters in diameter, located about 140 meters away from the lunar module model. It is not the Sun.

WAS IT SIMULATED IN A CAVERN, OR A DESERT?

If this is a cinematographic set, as the clues show, how was it done? Was it created inside a large cavern, far from prying eyes? Or in a private place in a desert far from urban centers? Or is there another option? What is quite evident is that the scene is not located on the Moon, specifically in the Sea of Tranquility. This conclusion is drawn from several discrepancies already indicated.

The answer as to whether somebody fabricated it outdoors or underground can be extrapolated from a flying bright object seen in the sky.

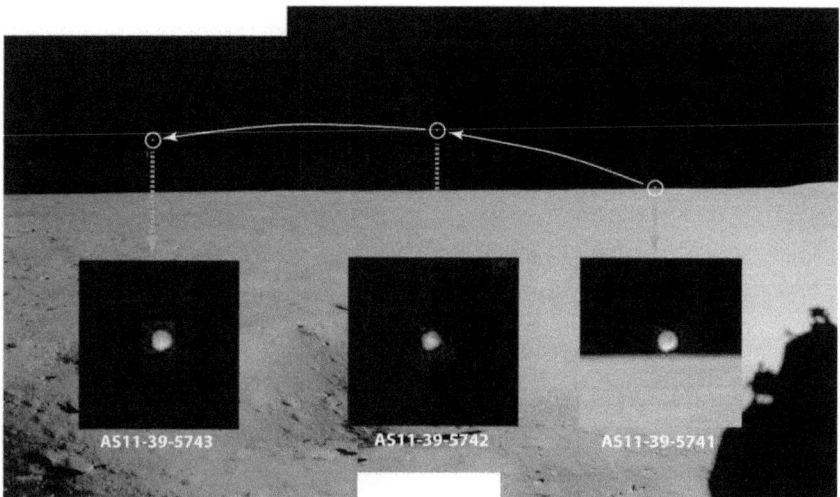

Figure 59 - Photo mosaic showing a luminous object moving in the sky.

In the same three photos we have shown before, with the "Double Crater" (which appears as a single one), we see a luminous point moving across the sky in each photo. It cannot be an internal reflection in the cabin's window from a light inside the spacecraft as the reflection moves backward. If the astronaut moved the camera to the left, as seen in this sequence, an internal reflection should shift to the right in the image, but it does not.

This bright object cannot be a piece of dust or a bug flying on the Moon.

Assume somebody still thinks that astronauts took these pictures on the Moon. Is that bright spot the Apollo 11 Command Module orbiting the Moon? In that case, it should be moving in a retrograde orbit from east to west. But this photographed object moves in reverse, from west to east. It is unlikely to be a UFO on the Moon, either. It is most likely an aircraft on Earth in a desert landscape.

In other photos I have analyzed, increasing the brightness of the dark sky revealed bright points that could be stars. It is an outdoor location on Earth, but not on the Moon.

So, this is a film studio set up in the desert. As demonstrated, it cannot be on the Moon. It resembles lunar settings, but it's distinctly different. Most likely, those who built this studio to simulate the Moon did it for training. They replicated the craters near the Lunar Module in a desert. They made small explosions and shaped the ground to resemble lunar soil with rocks and craters. They made the mistake of not digging thoroughly. One of the two craters of the "Double Crater" is missing, as seen in the photos, and perhaps they also forgot a crater in the central junction of the double crater. Another square crater is in the photos, but it is not on the Moon. They installed a crane or support 50 meters high and 140 meters away from the Lunar Module replica with very bright lamps covering a circle of 1.2 meters in diameter. They installed a fence or backdrop about 5 to 6 meters high toward one end of the landscape, facing the windows of the Lunar Module. On it, they painted what appeared to be a continuation of the desert landscape, crafting a horizon meant to mimic the Sea of Tranquility on the Moon. They took great pictures and black and white TV footage of the astronauts to transmit from the Moon, giving the impression that the Apollo 11 astronauts were walking on lunar soil, even though they were actually orbiting the Moon.

But with all this cinematographic effort, worthy of a great film director and supported by a reasonable budget, like all films, had some details that needed to be corrected. We have already pointed out some mistakes in this book.

As a conclusion in this chapter:

Fact #16: The analysis of the different clues in the Apollo 11 photos—such as the absence of some craters, the Lunar Module being displaced by 4 to 6 meters, three-dimensional images of the surrounding terrain suggesting the use of a backdrop on the horizon, and shadows indicating that the lighting originated from a cinematographic reflector rather than the Sun—leads to the conclusion that a cinematographic studio was likely constructed in a desert to mimic the lunar landscape. Photographers took good pictures there, and the video we observed on July 20, 1969.

APOLLO "Y" MISSION

We have established that the Apollo 11 astronauts could not have taken the photos on the Moon during their mission. The Sun looks too high. We reviewed several clues that the pictures were taken in a film studio, most likely in a desert area. The question is, who installed the equipment we see today on the Moon at the Apollo 11 landing site, if not Apollo 11? The answer is another Apollo mission.

On NASA's official YouTube channel, there is a video of Apollo 11, which NASA describes as a copy provided by CBS News and other sources of the original footage that was lost when NASA accidentally erased 14 magnetic tapes. That video shows activity on the Moon by alleged Apollo 11 astronauts. We have found that this video has two parts. The first is fake and absurd, where the astronauts walk vertically, but somebody cuts off the horizon with an 11-degree slope. The shadows cast by the Sun are horizontal, indicating they are not on a hill but on flat, horizontal ground. A video editor modified the first part of the video to show the horizon tilted, as it should be because the camera is supposed to be inclined. Did the original video have a flat horizon, and we didn't notice it when we watched the television broadcast in 1969?

The second part of the video shows that astronauts recorded it in a low-gravity environment with no atmosphere. The pendulum movement of the strap attached to an astronaut's chest, the way the grains scatter away from the ground, and the motion of the flag suggest that the footage was most likely recorded on the Moon and perhaps in the Sea of Tranquility. Real astronauts recorded this video during the Apollo "Y" Mission, which is the name I have given to that mysterious Apollo mission. For some reason, NASA did not include the first part of the video created by Apollo "Y." Perhaps the moment when the astronauts went down the ladder of the Lunar Module did not look good in their recording. Those who edited the video we see today decided not to include that part and left a clip from the pre-recorded studio version. They edited it but the result is peculiar.

So, which of the Apollo missions traveled to the Moon to install the equipment that Apollo 11 did not install? Was it a secret, additional mission unknown to us, or was it one of the recognized missions that traveled to the Moon?

If it had been an additional, unknown mission, launching a Saturn V rocket, it would have been difficult to go unnoticed. And it would have required sending the Navy to secretly pick up the astronauts in the ocean after their mission. And the astronauts' wives would have missed their husbands who embarked on a secret mission to the Moon. This hypothesis implies that more people were involved in the hoax, increasing the risk of exposure. So, then, it must have been one of the known Apollo missions.

Was it Apollo 12? Maybe not, given that Apollo 12 was the first mission to land on the Moon, proving that it was possible to do so. It must have been a later mission when the hoax proponents felt confident in successfully sending and landing a spacecraft on the Moon.

Was it one of the Apollo 14 to Apollo 17 missions, which successfully landed in different places on the Moon? To do so, they should have needed to carry significantly more fuel and a double set of equipment to install: their own mission's plus what was supposed to have been left by Apollo 11. Also, landing at their designated site, then taking off and landing in the Sea of Tranquility, required much more fuel and oxygen, and the mission would have taken much longer and perhaps would have been difficult to conceal. The fact that we currently see a lower part of a lunar module in the Apollo 11 landing site disqualifies any of these missions because there are also similar lunar module bases in the other Apollo 12 to 17 sites.

We have only one mission available, the most likely one that installed the Apollo 11 equipment without us realizing it: Apollo 13.

Almost everyone remembers Apollo 13. NASA aborted the mission after an oxygen tank explosion. They had the failure on their way to the Moon and returned without landing. There has even been a movie about that mission, and NASA, in general, won the sympathy of many people, seeing how fragile a space mission could be.

If that mission had installed the Apollo 11 equipment, we would have faced a second deception. Apollo 13 events led us to believe it was damaged, but it was probably not true.

Is it possible that Apollo 13 was able to land on the Moon without us noticing? For this to be possible, it must have been a speedy mission, with a swift journey to the Moon, landing quickly in the Sea of Tranquility, installing the equipment within a few hours, and then departing immediately without any break to sleep.

In the NASA YouTube channel video of the Apollo 11 extra-vehicular activity, which we know the second part is authentic and recorded on the Moon, the Sun is between 19 to 21.3 degrees elevation, as we have already estimated. Again, we have an average angle of 20 degrees. Did NASA need to land Apollo "Y" when the Sun was 20 degrees? Why didn't they choose 14 degrees as Apollo 11 should have been? The reason was to maintain consistency with existing photos, and in particular, that the shadows on the flag had to be identical between the video and the images that a photographer took in the recording studio. NASA published these photos in LIFE magazine and other publications of the time just after the arrival of the Apollo 11 astronauts. As we have already indicated, the flag looks starched, with the pre-established folds, which is why the photo in the studio and what we see in the Apollo "Y" video are somehow similar. However, we have already pointed out that the flag's shadows and shape do not match. If Apollo "Y" had landed with the Sun at a different elevation than 20 degrees, the differences in the shades on the flag would be too noticeable.

And is it possible that Apollo 13, if indeed it was the Apollo "Y" Mission, landed on the Moon when the Sun was at a 20-degree elevation? Yes, it would have had to travel quickly, arriving in 2 days or less. With JPL's Horizons[5] tool already mentioned, we can calculate the Sun's elevation to check on what date and time the Sun was at 20 degrees elevation. With that tool, we conclude that if Apollo 13 would arrive in 2 days or less, the Sun would be at 20 degrees of elevation.

Apollo 8, not wanting to break a speed record, reached

the Moon in 2 days and 19 hours. So, Apollo 13 could have arrived in less than two days (under 48 hours), spent a few hours landing and setting up all the equipment, and returned without haste.

We know that the Apollo 13 astronauts had a conference with their families around 55 hours into the mission, and a few minutes later, the failure occurred. It could have happened when they returned, not on their way to the Moon.

The Apollo spacecraft could travel to the Moon quickly or slowly. That depended on the spacecraft's mass and, above all, the fuel it uses. To reach the Moon, stage three of the rocket boosted the Service Module, the Command Module, and the Lunar Module for about 6 minutes towards the Moon, leaving the Earth's orbit. If the rocket ignited longer, the escape velocity from the Earth's gravitational field would be higher. Therefore, it would go faster and arrive in less time. When approaching the Moon, the craft must consume more fuel than usual to slow down the fast pace of the spacecraft. We see that the Apollo missions varied in the time they took to reach the Moon. That's because of the planned fuel consumption. So yes, Apollo 13 could have gone to the Moon very fast, using much fuel on its outbound trip and saving it on its return trip.

Interestingly, Apollo 13 carried an additional 22,000 pounds of weight[27], primarily fuel loaded into the tanks. According to NASA, they did it to test for future missions when they would take more cargo. The craft may have used that extra fuel to get to the Moon faster, not for testing.

How did they fake the damage? Did they unload the oxygen from a tank and pretend they heard an explosion, which caused the loss of that oxygen tank? If they had caused an explosion on purpose, they would have risked themselves while still inside the spacecraft. It was safer to cause the explosion just before descending to Earth, when they were out of the service module and getting ready to plunge into the ocean to be picked up by the Navy.

There are 22 photos taken with one of the Hasselblad cameras on board, shot by one of the astronauts through the spacecraft window, showing the service module damaged by the explosion.

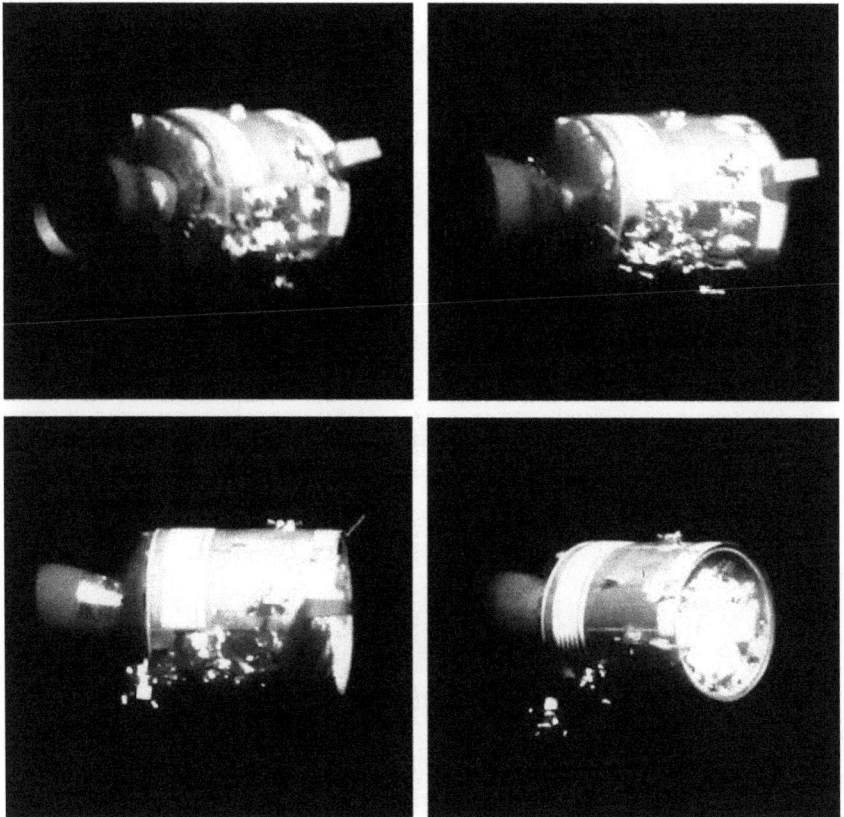

Figure 60 - Damage to Apollo 13. Photos taken when the astronauts undocked from the explosion-damaged service module. Why does it rotate on two axes?

The Apollo 13 astronauts took these pictures from the capsule after undocking from the service module and just before returning to Earth. Some people have created animated videos with the photos, such as the one on the Apollo Remastered[28] YouTube channel, where we see the service module rotating and with one side damaged by the explosion (see Figure 60).

Looking at the animation and the different shots and observing the shadows, we notice that the module rotates on two axes. The spacecraft is normally rotating on its longitudinal axis to avoid overheating on one side by solar radiation. Still, the rotation on the other axis does not seem natural. The astronauts could have caused the explosion on purpose, but they were no longer in the service module. They might use a remote detonation or a timer. That caused the module to rotate on the other axis.

If the explosion had occurred before, when the astronauts announced they had a problem, the spacecraft would have pivoted. They would have had to stabilize it to continue the trip, rotating in its longitudinal axis only. So, on the return, when undocking from the service module to enter Earth's atmosphere, the service module should not rotate in the manner observed in the 22 photos, on two axes, unless a significant external force caused it, like a detonation.

The hypothesis is that they did cause the explosion on purpose when they were already about to descend to Earth and outside the service module. They took several pictures as evidence of the damage caused.

In another chapter of this book, we mentioned that the location of the Apollo 11 Lunar Module in recent photos of the Moon does not match the area we see in the images, suggesting that a photographer took them in a recording studio. Apollo "Y" landed 4 to 6 meters (13 to 20 feet) east of where it should have landed. Is such a precision landing possible?

The Apollo missions evolved. Apollo 11, designated "G" Class, had poor landing accuracy. NASA records[31] state that it had a landing error of 6780 meters (22,244 feet). NASA claims it landed almost 7 kilometers from its target. That level of accuracy was acceptable for Apollo 11, but later missions required greater precision, especially when landing in hilly areas. Apollo 12 was the first "H" Class mission[31], demonstrating that it could land accurately. Apollo 12 descended to within 163 meters (535 feet) of its target. This landing offset presents a significant improvement over the accuracy achievable by Apollo 11. Apollo 14 was accurate to within 53 meters (174 feet) of its target.

So, if Apollo "Y" was Apollo 13, having an accuracy of 4 to 6 meters doesn't sound far-fetched. We should consider that just a few meters above the lunar surface, the astronauts had some maneuverability to adjust their landing point by a few meters, for example, to avoid craters and look for a site free of rocks. They did not necessarily have to land at a designed point, as they could decide at the last moment to land the lunar module at a stable and safe site. They could then approach the target precisely if that became necessary. Apollo 12 sought a landing site on the edge of a large crater to walk to the Surveyor 3 probe on the opposite edge of the crater. Apollo 13 landed with an accuracy of 4 to 6 meters because it had a clear motivation: to land precisely next to the Double Crater. They could not descend the spacecraft to the exact landing site, but it was an incredible feat to be close to it, just a few meters away.

Watching the NASA video of the "Apollo 11" landing[12] recorded with a 16mm film camera located on the right window of the lunar module, we notice that in the last seconds, the astronaut maneuvering the craft moved to the left to get closer to the Double Crater. Why did he do that? Landing farther away from the Double Crater was safer, but not very close to it. The reason for doing it might be to land at the designated spot, the same one used in the recording studio.

That Apollo 13 was the mysterious Apollo "Y" Mission is a hypothesis. I do not have enough facts so far to confirm it. The door is open for other researchers who know how the command and lunar module controls work and how to alter the telemetry. I know that Apollo 13 was carrying more fuel than was needed for a timely arrival on the Moon, a landing in the Sea of Tranquility, and the installation of Apollo 11's instruments. And I know the damaged service module was rotating on two axes, not one. More evidence is required.

If Apollo 13 was indeed the mission that installed all the Apollo 11 equipment, we should congratulate NASA for a flawless mission that arrived in record time to the Moon, landed very close to the designated point, installed the missing Apollo 11 equipment, and managed to fake a spacecraft failure, fooling us all. It was a real and effective mission, but it was a lie. And Apollo 11 was another lie, also not real. With an accuracy of 7 kilometers at the landing point, it is understandable that some people at NASA, or those behind the fraud, lacked confidence in a successful Moon landing. There was a real risk of landing on the edge of a crater, hitting a rock, or rolling over.

The "Lunar Lander" probe recently stumbled and flipped on its side while attempting to land on the Moon. If this had happened to one of the Apollo missions, its astronauts would indeed have died there. It would have been prudent to wait for the certainty of a more accurate landing, as was Apollo 12, the first mission to land astronauts on the lunar surface.

Someone might think the Apollo 13 astronauts, who are still alive, might tell the truth. But, if someone hypnotized them, like the Apollo 11 astronauts, it would be difficult for them to remember the facts. Scientific studies[29] have shown that false memories can be implanted in people. If it is done under hypnosis, it is easier to do so. And suppose the astronauts were practicing several times all the activities they would perform on the Moon, in the simulators, training centers, or even in the movie studio. In that case, it is easier for them to take those memories of those tests as real and think they happened on the Moon.

The Apollo 11 and Apollo 13 astronauts, in my opinion, were victims, as were all of humanity who have been deceived.

HOW IT ALL HAPPENED

So far, I have presented several pieces of evidence that establish facts anyone can verify. No knowledge of Aerospace Engineering is required to confirm them. A high school student, or almost anyone with a curious and critical spirit, can find the truth by analyzing the available evidence in the Apollo 11 photos and video.

This chapter will present various hypotheses of what could have happened, which may explain these findings. The reader will be able to develop their own ideas as I present possible events that piece this puzzle together.

Today, many people declare the Apollo 11 issue closed. They believe that all those conspiracy theories have no basis in showing definitive proof that Apollo 11 did land on the Moon:

Fact #2: Several photos taken by the Lunar Reconnaissance probe in 2009 show Apollo 11 equipment installed on the Moon, indicating that some spacecraft put it there.

And if I try to show evidence that Apollo 11 never landed based on various analyses performed, for example:

Fact #3: The Sun is 6 to 7 degrees higher in the Apollo 11 Mission photos than it should be. The Apollo 11 astronauts could not have taken the Apollo 11 pictures on the Moon.

That immediately causes rejection, and I am labeled a "Moon Hoaxer." Administrators on scientific forums where I ask questions about the Apollo program reject me and delete my comments. Even in Facebook groups, they delete my comments and evidence when I present stereoscopic images that indicate a backdrop to the lunar landscape.

Many people are tired of conspiracy theories and do not want to know more about Apollo 11, considering there is too much evidence to prove that NASA reached the Moon. The amount of misinformation creates an armor in people's minds to accept facts that seek to get to the truth. However, suppose humanity plans to return to the Moon. In that case, we must know the truth. This enables us to learn from our past mistakes and lies so we do not fall into the same trap again. Recognizing the facts not only gives credibility to our endeavors but also allows us to learn from them, even if the truth is uncomfortable to acknowledge.

So, what happened with Apollo 11? I will present several hypotheses that might explain what occurred. I recommend taking these hypotheses as possibilities rather than definitive truths.

BACKGROUND

As we have mentioned, Apollo 8 was the first Apollo mission to reach the Moon. It did so in a record time of 2 days and 20 hours. The Apollo 10 mission flew close to the lunar surface, only 14 kilometers above it, and captured images, some stereoscopic, of the terrain where Apollo 11 was to land. However, the Apollo 11 mission was "G" class, whose lunar module and landing procedures on the Moon were not precise. NASA reported that it landed 6780 meters from its assigned landing zone (22,550 feet). We know it did not land, but that value indicates the level of error they expected to obtain.

Landing a spacecraft using Apollo 11's technology needed to be more accurate. NASA planned enhancements to improve accuracy, but these were not implemented until Apollo 12, and there was no assurance they would be effective.

I hypothesize that perhaps a political figure, rather than a scientist, assessed the risk of the first Moon landing attempt potentially failing and chose to fake the landing. Scientists are usually fearless by failure. Every failure teaches and brings us closer to the truth. But a politician is different. Perhaps it was Richard Nixon himself who pushed this deception, supported by Wernher von Braun and a small group of people involved, who planned the way to fake a landing. Neither of them is alive today, so we cannot ask them to confirm or deny it. There was a lot of political pressure and a space race competition with the Soviet Union that may have created pressure to go down that path of lies.

THE PREPARATION OF THE DECEPTION

Thus, the people who participated in the simulation decided to make a replica of the landing site in a desert. They used photos taken by Apollo 10 and built the fake lunar landscape. They made explosions in the ground to create craters and removed material and rocks to recreate the replica of the Sea of Tranquility near the Little West and Double Craters. They installed a duplicate of the Lunar Module there, and about 200 meters away, in front of the spacecraft's windows, installed a fence that an artist painted to simulate the continuation of the surrounding terrain displaying a fake lunar horizon. The backdrop would be at least 6 meters high, the same height as the Lunar Module.

Fact #14: Stereoscopic analysis of Apollo 11 photos indicates the presence of a distant screen that simulates the lunar landscape.

However, the construction of the craters could have been better executed. The excavator failed to make some craters and misplaced others. After all, the Apollo 10 photos could not possibly show excellent detail, but they did their best with the information available.

120

Fact #12: In the NASA Apollo 11 pictures, we see significant differences in the craters around the base of the Lunar Module towards the southwest, comparing them against images from the Lunar Reconnaissance probe.

They took the photos at night. To simulate sunlight, they installed one or several cinematic reflectors, forming a shining circle of 1.2 meters in diameter (4 feet). And somebody chose an elevation of the "artificial sun" of 20 degrees as a design parameter. For this purpose, the operators installed the reflector on a scaffold, support, or crane, which raised it to a height of about 50 meters. The elevation could be changed if necessary, but moving the crane forward or backward was challenging in that terrain.

Fact #15: Analysis of the shadows of the Apollo 11 photos, in their penumbra dimension, shows us that what illuminates the landscape and the astronauts is a cinematographic reflector, perhaps 1.2 meters in diameter, located about 140 meters away from the lunar module model. It is not the Sun.

Then, before the trip, the crew conducted many tests of the activities they were scheduled to perform. Perhaps at the beginning, the astronauts did not know that what they were doing there would be used to depict a fictitious landing. They likely viewed it as just another routine part of their training.

The cinematographers captured footage of the Moon's surface descent, the astronauts stepping down the Lunar Module's stairs, and the equipment's installation. They took photos with Hasselblad cameras, similar to the ones the astronauts would carry on their trip to the Moon, capturing both photos and several videos.

The studio crew made a few mistakes while recording the videos. In Armstrong and Aldrin's descent down the spacecraft's ladder, they should have installed the TV camera tilted. The horizon in the final movie was horizontal, not with an 11-degree inclination. Additionally, they recorded it with the reflector in a higher position, about 30 degrees. The current video on NASA's official site is a copy provided by CBS News, as the 14 tapes of the telemetry transmitted by Apollo 11 disappeared, and NASA believes they accidentally erased them.

Fact #4: The actual video of the Apollo 11 mission with its activity on the Lunar surface is not the original. It is a recent copy. The original no longer exists.

In the initial part of the video copy, it is clear that an editor altered the lunar horizon profile to make it look tilted, but the astronauts walk vertically. Editing the video to show the astronauts walking at an angle would have been challenging, or they did not consider this additional correction necessary. We do not see the image inverted at any time, even when the astronauts move the camera from an inverted position to a normal one.

Fact #6: Even though the TV camera starts upside down and rights itself when taken by the astronauts, we do not see in the NASA video that the Apollo 11 image is inverted at any time.

As a result of all the video recordings and photos taken, they had the necessary material to carry out the deception. A few days before, the decision not to land the spacecraft was communicated to the astronauts, and they were asked to collaborate to participate in this lie. We do not know what pressures or motivations led them to participate. For sure, somebody forced them.

THE TRIP

The main requirement was to involve as few people as possible in the deception. The astronauts needed to board the spacecraft and embark on their journey. It is possible that a hoax, known to only a select few, could be carried out while the astronauts traveled in the spacecraft. They boarded the rocket, took off, and met the Navy after splashing in the ocean on their return. Additionally, astronauts must transmit from the Moon, not necessarily on the Moon's surface, to the ground stations. Many stations worldwide collaborated with NASA to receive the transmission and telemetry from the spacecraft. Sending messages from Earth rather than from the Moon or failing to return in a capsule would have immediately exposed the fraud. Therefore, they did indeed travel to the Moon.

After the Apollo 11 liftoff, they followed the same trajectory as Apollo 10. They arrived on the Moon and descended in the Lunar Module as Apollo 10 did, but they did not touch the lunar soil. They could have fallen into a crater, or the descent system might not have worked correctly, or they tipped over despite the landing tests done on Earth. If the ship had flipped over, rescuing them would have been impossible, as the oxygen would have run out very soon if they had survived the fall.

During the descent, they switched the live feed to a recorded feed. The astronauts from the Lunar Module began transmitting the pre-recorded video in the telemetry system and TV signal. Supposedly, they landed, but it was not true; Aldrin and Armstrong were still in orbit around the Moon, perhaps in the Lunar Module, or reunited with Collins in the Service Module.

Stations in California and Australia received the signal. However, part of the simulation did not include installing the external antenna. They transmitted everything using the small antenna on the Lunar Module roof. The plan and the pre-recordings did not include installing the umbrella-like and sophisticated antenna since it would be better to send a poor-quality signal, which would be challenging to review and compare with photos of future lunar missions.

Fact #5: The Apollo 11 astronauts carried an antenna to transmit a better-quality television signal but decided not to install it. That led to receiving a poor-quality signal sent from the Moon.

That external antenna was sophisticated, constructed of 38 miles of gold wire[17], and looked like a large umbrella. NASA reported they carried it on board but did not install it. Later Apollo missions did use that external antenna. The justification was that it would take 19 minutes off the whole process. They had an additional 12 hours after the extra-vehicular activity, which included 5 hours of sleep. They would have had plenty of time to install the antenna if they walked on the Moon. Still, it was better not to install it, and we know the real reason for the transmission not showing a higher resolution. The pre-recordings made on Earth did not include the big antenna installation.

For some reason, they decided to start the simulation of the first moonwalk earlier. As chronicled in Science + Media Musseum[17]: "After successfully landing on the Moon, Neil Armstrong and Buzz Aldrin were due for a few hours of sleep. They couldn't wait and requested to leave the lunar module earlier than planned." So they left early, or rather, they began transmitting the pre-recorded video earlier than planned. This decision likely had to do with the availability of antennas on Earth ready to receive the signals from the Moon. Or, since they were not on the surface but flying around the Moon, they might lose communication later as they passed behind the Moon. For some reason, they started the simulation of the moonwalk earlier while the Sun was still very low, near the horizon, at about 14 degrees elevation. This means the scam would have the discrepancy that the photos previously taken in the film studio featured studio lights set at 20-degree elevation. They were supposed to be on the lunar surface with the Sun at 20-degree elevation, not 14 degrees. Perhaps they thought that was a minor detail that no one would notice.

The astronauts transmitted the voice signal live. The video broadcasted was fake, but the dialog was in real-time. That way, they could interact with the mission operators in Houston or even talk to President Nixon. The famous phrase "That's one small step for man, one giant leap for mankind" was declared by Armstrong while he was flying around the Moon, not standing on the lunar surface.

THE RETURN

When the astronauts returned from the Moon, NASA confined them to quarantine for several days, isolated from the world. Perhaps at this time, they underwent hypnosis. The scammers had to ensure astronauts remembered that they had walked on the Moon. The hypnosis must have focused on recalling the details of their activities in the simulators and the film studio.

After quarantine, they traveled the world like heroes before the Apollo 12 trip. They gave away rock samples, supposedly collected during their journey to the Moon. But, the reader may ask, if they did not travel, where did they get those moon rocks? Interestingly, the Dutch National Museum, Rijksmuseum[30], reported that a "moon rock" they had, which was donated by somebody received from those Apollo 11 astronauts during their tour, was a piece of a petrified tree. In other words, an earth rock. It would be interesting if owners of moon rocks from the Apollo 11 tour conducted tests to verify the authenticity.

After the Apollo 11 astronauts returned, NASA sent some photos to the media for publication. One of them shows Aldrin saluting the U.S. flag.

Fact #11: NASA has retouched some of the Apollo 11 photos.

I have pointed out that one of the photos of Aldrin saluting the flag is a fake. Nobody even took it in the film studio or on the Moon. That picture is a copy of another photo from the studio that somebody edited.

Several of the Apollo 11 photos we currently see in NASA records, including the images with the U.S. flag, show that the Sun was higher than it should have been.

Fact #3: The Sun is 6 to 7 degrees higher in the Apollo 11 Mission photos than it should be. The Apollo 11 astronauts could not have taken the Apollo 11 pictures on the Moon.

The photos show an elevation between 20 to 21 degrees, while it should be 14 to 15 degrees. That equals a difference of 12 hours, which makes it impossible for the astronauts to have taken them.

At that time, the photos were spectacular. Many people believed that this extraordinary trip had occurred.

THE APOLLO 12

Time passed, and along came the Apollo 12 mission. It was the first human-crewed mission to land on the Moon. It touched down near the Surveyor 3 probe, allowing the astronauts to walk up to it.

Those who performed the hoax knew that the "H" class missions might land on the Moon and do so with reasonable accuracy. Among the possibilities considered, there was concern that Apollo 12 might fail in its attempt. However, if Apollo 12 were not successful, it would not have been so embarrassing since they had already pretended that Apollo 11 had succeeded in landing, and the space program would not have stopped.

PREPARATIONS FOR APOLLO 13

Then, the second phase of the plan began. Everything was to be prepared on Apollo 13 so that it would carry the Apollo 11 instruments to the Moon. They knew that in the future, with better telescopes and new spacecraft reaching the

Moon, if other nations discovered that there was nothing at the Apollo 11 landing site, the deception would be blatant. They had to install the Apollo 11 equipment.

The small team of scammers made the entire flight plan, calculated the amount of fuel required, checked the activities to perform to simulate the spacecraft's failure, and made detailed simulations to repeat the script of the activities that the Apollo 11 astronauts were supposed to do on the Moon. The timing and sequences of what they were to do were to be repeated with meticulous care, as they planned to record a video and take some pictures on the Moon.

APOLLO 13'S JOURNEY

Apollo 13 lifted off on April 11, 1970. It carried more fuel than it would typically require. The spacecraft ascended, and the second stage was activated, lifting it into Earth's orbit. It brought the same equipment on board as Apollo 11.

In Earth's orbit, they ignited the third stage until they exceeded the speed planned for their original mission. They were now on track to get to the Moon in less than two days.

There was a period of silence where they were supposed to sleep, but in reality, they were landing on the Moon. They descended carefully, approaching Little West Crater, and slowly plunged towards the Double Crater. They did their best and landed 4 to 6 meters from the indicated spot. It required that in the last seconds of descent, the pilot moved the lunar module closer to the Double Crater, as the Apollo 11 descent video shows, which was recorded in the Apollo 13 mission. That was close enough. This stage of the plan was successful.

They quickly went out to perform the same activities that Apollo 11 had planned. For some reason, the video recording they made of them coming down the stairs of the Lunar Module did not look good. Perhaps something didn't seem like the original transmission, and later, the video editors would remove that part of the recording.

They collected rock samples that somebody would later replace with the fake moon rocks in the NASA archives. They pulled out the U.S. flag and carefully unrolled it. The flag was starched and had a previously designed shape and undulations. It had to be identical to the flag they used in the staged photos in the film studio. The flag oscillated due to inertia but lacked the drag typically seen in an atmospheric environment.

Fact #1: The flag movement in the NASA videos, such as Apollo 11 and Apollo 17, indicates that it occurred in an atmosphere-less environment, such as the Moon.

They arranged the flag so that its horizontal crossbar at the top pointed to the Sun. They moved it several times until the shadows matched the programmed design. Although they were very similar, they were not identical.

Fact #10: When comparing the photos of the NASA records on Apollo 11, where they show the US flag, with the video on NASA's YouTube channel, we see differences in shape and shadows. They are similar flags, but they are not the same flag.

"Aldrin" stood in the defined position near the flag and saluted it while "Armstrong" took pictures of him. The ribbon hanging from the chest of "Aldrin's" suit, or the Apollo 13 astronaut who replaced him, moved in a pendulum motion for several seconds without stopping.

Fact #8: The pendulum motion of the tape hanging from the astronauts' suits, specifically its oscillation period, and how it is damped, indicates that the first part of the video was recorded in a terrestrial environment with an atmosphere, and the second part in a lunar environment with low gravity, and no atmosphere.

They moved around the area, setting up instruments and sometimes making small jumps. They took more pictures and left the TV camera recording the entire 2-hour sequence of extra-vehicular activity.

They climbed back into the spacecraft and, without wasting time, filled it again with oxygen and prepared to return. There was no time to sleep.

They took off and rendezvoused with the third astronaut, who was awaiting them in the Command and Service Module while orbiting the Moon. They brought photos and videos taken on the Moon, and authentic Moon rocks. Then, they began their return to Earth.

For the operators in Houston, the astronauts were just waking up from over 5 hours of sleep, still en route to the Moon, not returning from it. The astronauts conducted routine activities. At 55 hours into the trip, about 7 hours after landing on the Moon, the Apollo 13 astronauts, who had not yet been able to sleep, teleconferenced with their wives in Houston. The wives thought the astronauts were on their way to the Moon, but they had already been there.

A few minutes after the conference with their families, the astronauts unloaded the oxygen from one of the tanks. They cut off communication for a couple of seconds, then re-established it. And they reported that they had a problem. They said they heard an explosion, which didn't happen. The telemetry showed several alarms, and the oxygen level in one of the tanks dropped completely. From then on, it was demanding and stressful work for the ground engineers, along with the astronauts' families and countless others who followed the events closely. They were deeply concerned about the fate of Apollo 13. Contrary to the alarms, the astronauts remained calm because everything was under control, and they knew the oxygen they carried would not run out. There had been no explosion. Everything was proceeding exactly as planned.

Approaching Earth, they detached the capsule from the service module. They remotely activated an explosive device pre-installed on the side of the service module. A few minutes after the explosion, as the debris cleared, the Apollo astronauts took several photos from the capsule window of the damaged service module, which was rotating because of the blast.

The Apollo 13 astronauts finally descended into the ocean, and the U.S. Navy picked them up.

POST-ARRIVAL ACTIVITIES OF APOLLO 13

The astronauts were subjected to several interrogations. They agreed to receive hypnosis, which was used to implant false memories, convincing them that the mission indeed had been a failure but that they managed to return safely.

Several press conferences were held, and investigations

were conducted to determine what happened. Some journalists asked several questions[27], like why they were carrying 20,000 pounds of excess weight, specifically whether it was fuel, and where exactly had this fuel been loaded on the ship.

The conspirators undertook various actions to cover up the hoax. They replaced the Moon rocks the Apollo 13 astronauts brought back. They analyzed the photos and videos obtained by the astronauts. They might have mixed some pictures with current Apollo 11 fake photos in NASA's records. Still, the conspirators could not replace those images near the Lunar Module that clearly showed the craters from the studio's artificial lunar landscape. The scammers could not replace all those Apollo 11 photos, such as those showing the U.S. flag, Aldrin's descent down the stairs, and others already published.

At some point, they destroyed all 14 tapes of the original Apollo 11 Telemetry sent by Apollo 11 on July 20, 1969.

Specialized editors of the conspiracy group analyzed the video taken by Apollo 13 on the Moon. They decided not to include the first part because it did not look good, and there was no way to repeat the shot. And they noticed an error in the original video, which NASA showed in the 1969 transmission. The horizon when Armstrong and Aldrin descended on the ladder was horizontal, not tilted. They compared it with the Apollo 12 videos. They realized the image must be inverted and the horizon inclined 11 degrees. So they edited the original 1969 video, cropping the horizon to make it appear tilted. They could not change the vertical position of the astronauts or their horizontal shadows, but they hoped nobody would notice it. They made the transition in the video, between the first part from the original edited film and the new Apollo 13 video, showing white frames as if the astronauts were moving the camera. What we see today is a combination of two different videos.

Fact #7: The NASA Apollo 11 video shows two very dissimilar environments, indicating that it comes from two different shots taken in separate locations and with different lenses. The elevation of the Sun in the first part is too high, about 34 degrees. In comparison, the second part of the video shows the height of the Sun at about 20 degrees, similar to that of the NASA photos. Neither matches the expected value of 14 degrees from the Apollo 11 Mission.

In short, someone edited the Apollo 11 video we see today on the NASA YouTube channel. They included the first part from the original 1969 fake video, and the second part has the film the Apollo 13 astronauts recorded on the Moon. They expect no one will notice the difference.

Fact #9: The NASA video illustrates the different effects of boots hitting the ground. The first part suggests that it was recorded in a terrestrial environment with an atmosphere, and the second part that it was recorded on the Moon.

In the second part of the video, they may have added some small rocks that match those of the fake photos taken in the film studio. This effect is easy to achieve on the video recorded from a static tripod using post-editing video tools.

NASA claimed they received a copy from CBS News archives and other sources to post an edited version on their YouTube channel as they realized the original Apollo 11 tapes had disappeared.

LUNAR RECONNAISSANCE MISSION

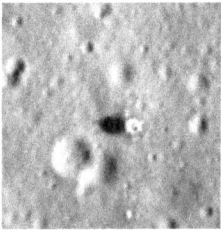 Years later, in 2009, NASA sent the Lunar Reconnaissance probe to the Moon. The photos this craft took show the Apollo missions' landing sites. The Apollo 11 equipment, like other Apollo missions, is in the images. This is definitive proof that NASA did reach the Moon, that all conspiracy theories are unfounded, and that all suspicions are resolved. Case closed: Apollo 11 did land on the Moon.

Is everything proven after the photos of this probe, and there is no room for discussion about it? Well, no, these photos present new doubts.

Fact #13: The location of the Lunar Module varies by 4 to 6 meters. Its position on the Moon's surface today differs from what the Apollo 11 photos show.

Some aspects of the Lunar Reconnaissance photos do not match. What we see on the Moon today differs from what the Apollo 11 photos show. Somebody shot the pictures in a cinematographic studio.

Fact #16: The analysis of the different clues in the Apollo 11 photos—such as the absence of some craters, the Lunar Module being displaced by 4 to 6 meters, three-dimensional images of the surrounding terrain suggesting the use of a backdrop on the horizon, and shadows indicating that the lighting originated from a cinematographic reflector rather than the Sun—leads to the conclusion that a cinematographic studio was likely constructed in a desert to mimic the lunar landscape. Photographers took good pictures there, and the video we observed on July 20, 1969.

These facts reopen the Apollo 11 case!

You might find this explanation very imaginative, like a Hollywood movie. In reality, the true events are likely even more fantastic, and this is just a simplified version of what happened.

Future missions to the Moon, specifically to the Sea of Tranquility, will re-examine details like the rocks near the Lunar Module, the astronauts' footprints, and small craters. These investigations aim to confirm whether what is on the Moon today is not what the Apollo 11 photos show. The deception will eventually be too apparent, and someone will have to tell the truth about what happened. Perhaps it is time to bring all these facts to light.

HIDING LIES

How can you hide a lie? By mixing it with several truths. Two lies have been hidden here: one about Apollo 11, which did not land on the Moon, and we were led to believe that it did, and the second is that Apollo 13's mission, which went to install the equipment that Apollo 11 failed to, despite the claims that it encountered a critical failure.

I get contrary responses when I share the facts presented in this book. It is human nature to move to extremes. And as long as extreme, irreconcilable positions cling to their ideas without wanting to change, the truth, which likely lies somewhere in between, cannot be found. It is as if most people can only see the world in black and white without recognizing that there are shades of gray or various colors. Some say, "It's either black or white; NASA either lied or didn't lie; there is no in-between." However, they fail to recognize that a tiny group infiltrated NASA and lied, not NASA as an institution. Some people who work or have worked at NASA most likely do not know about the Apollo 11 and Apollo 13 deceptions. It's more complex than saying NASA lied.

By showing this evidence in social networks, those who

think all Apollo missions are fake emerge. They believe that NASA never made it to the Moon. Perhaps they discovered some evidence that made them see that Apollo 11 was a hoax, and now they consider that everything has been a hoax. In this group, there are varying degrees of skepticism, where some, despite seeing the evidence that NASA did make successful space trips to the Moon, believe they are being deceived and will not admit the truth. In some extreme cases, members of this group refuse to acknowledge evidence even when it's directly in front of them and see deception in everything they encounter. Some even believe that the Earth is flat or that we live in the Matrix, and everything is false and unreal. I was trying to contact someone skeptical of NASA's space travel to the Moon, but the response I got was to suggest that aliens are demons and that the Earth may be flat.

The other extreme includes those who refuse to accept objective evidence or facts suggesting that Apollo 11 never landed on the Moon. Upon contacting them, they unfairly label me a "Moon Hoaxer"! They put me in an extreme position, which they despised for lacking scientific rigor, not knowing I was not part of that group. But, in their own skepticism, they fail to acknowledge reality themselves.

I have found that the truth lies somewhere in between. And that is what this book is about, showing that it is not one extreme or the other but somewhere in between. I have pointed out that NASA successfully landed men on the Moon. I have confirmed that they have stepped on lunar soil by verifying how the straps that the astronauts wear strapped to their chests move in a pendulum-like manner, where they demonstrate that it is a low-gravity environment. Or by accidentally kicking the lunar soil and moving it long distances, indicating a low-gravity environment with no atmosphere. Or by showing photos from recent missions, such as the Lunar Reconnaissance, which proves that an Apollo spacecraft did land on the Moon. These facts support people in one extreme, indicating that trips to the Moon did occur. However, some, to the extreme, do not accept this truth.

And I have found that Apollo 11 did not land. Armstrong and Aldrin could not have taken the pictures we see of the Apollo 11 mission. They went out supposedly to walk on the lunar soil and set up the equipment when the Sun was very low. They could not have taken those pictures because the positioning of the Sun was higher in the Apollo 11 pictures, and by the time the equipment was installed there, the astronauts should have been preparing to return. This fact is something that anyone can prove to themselves. And I have shown that what we see in those Apollo 11 photos does not match what is on the Moon. Some craters are different, and the lunar module is displaced. The stereoscopic images indicate a backdrop, and the lights and shadows in its penumbra are not due to the Sun but to a cinematographic lamp. All this shows the other extreme of a hoax using photos taken in a recording studio. But some do not accept these facts.

Both extremes are right and wrong. The truth is that NASA did get to the Moon, but they lied to us. So it's not black and white. It is something in between.

How do you keep such a deception hidden? By showing evidence that satisfies every extreme position and fuels the controversy. Whenever a skeptic presents evidence suggesting that the Moon landings were faked, even if the evidence is lacking, the other extreme immediately emerges and firmly rejects such a possibility. That creates extreme viewpoints, where one side is unwilling to consider the evidence of fraud, while the other refuses to see proves that lunar travel was authentic. It is a circle of disinformation that goes back and forth, blinding those participating in this game of extreme positions to open their eyes and analyze the truth. They can see that the evidence is compelling, but they don't want to admit that they have been wrong in their extreme position and don't want to move to the middle ground to find the truth.

People who have settled in those extreme positions will probably not like this book. I have not written it to agree with one group or the other. I have written it to share the truth I have found and to tell these people who have been in those extreme positions that there is a middle ground, showing tangible evidence of lies and truths.

It terrifies me to see humanity today being influenced and manipulated by what social media presents. They base their opinions on what they see on the internet, not what they observe in nature. Many must use their knowledge and common sense to seek the truth. Those who want to manipulate humanity easily create conspiracy theories full of false information, like the Flat Earth theory, or blur those conspiracy theories that do have a real foundation, like the Apollo 11 hoax. These orchestrators saturate skeptics with disinformation and false evidence to drive the two groups to each extreme. They give each group the fodder they need to stay in their extreme position and avoid the middle ground, making it challenging to find the truth.

Seeing people who prefer to live in illusions and lies terrifies me. They live in a fairy tale where everything is heroic and extraordinary or in a horror story where the bad guys want to deceive humanity. Reality is neither a fairy tale nor a horror story.

We are in the age of Phantasmagoria, where we can no longer distinguish reality from fiction. We are lied to and accept it; we are deceived and take it. And so do politicians, religious leaders, and the media. There is so much information and misinformation that it is difficult for some people to recognize the truth.

And how can we find the truth without being manipulated? By being skeptical and open-minded at the same time and observing nature more than what a TikTok or YouTube video shows. We need to find a neutral standpoint from which to observe the truth objectively. We must resist the temptation to promote disinformation for personal gain and help guide humanity back to a factual reality. We must search within ourselves for those answers. The universe is a reservoir of wisdom; we must be relaxed, neutral, and positive to understand and observe the answers.

Why is it important to recognize these deceptions and lies? Some tell me that exposing these lies is unnecessary because NASA has successfully taken humans to the Moon multiple times. It is essential to point out that lying is a mistake, and if we recognize our mistakes, we will learn and evolve.

Unfortunately, human society has moved away from some natural laws. We are in a process of evolution, perfecting ourselves a little more each day. Therefore, making mistakes is part of a natural process, which allows us to improve each time. Making mistakes is natural, but not recognizing and learning from them is a mistake. If we want to return to the Moon or slowly advance towards the stars, we must learn from our mistakes and improve. A building constructed on lies will collapse sooner or later. We must lay the foundation of truth in our space adventure.

There is a misconception in humanity about perfection and leadership. There is no such thing as perfection. Projecting an image of flawless leadership and perfection only leads to an inevitable destination: lying to appear perfect. Lying is a mistake many people make, especially politicians and religious leaders. No one is perfect or a saint. We are just people who make mistakes and can learn from them.

I still look at the Moon at night and dream of traveling there. And I gaze at the stars and imagine the countless civilizations that, like us, make the same mistakes and evolve in this vast, beautiful universe. I wish I could travel to the stars, explore new worlds, and learn from the experiences of other cultures.

When I see the chaos in our world—how we mistreat it, compete, and create conflicts among all human beings, ignoring that we are all interconnected and evolving together—I realize that we still have a lot to learn and discover before we can travel through space in search of new adventures. But we must begin the journey by recognizing our limitations and mistakes. The road will be complex, but it promises to be fascinating.

NOTES

1. The Future of Mankind - A Billy Meier Wiki - The Pleiadian/Plejaren Contact Reports. (n.d.). Www.futureofmankind.co.uk. Retrieved March 15, 2024, from https://www.futureofmankind.co.uk/Billy_Meier/The_Pleiadian/Plejaren_Contact_Reports#Indexed_0-9_A-Z_by_specific_information_and_where_to_find_it.

    See Contacts #203, 214, 231, 238, 357, 398, 433, 527, 529, 670, 672, 688, 709, 757, 770, 775 and 779

2. Horn, M. (n.d.). Video [Review of Video]. They Fly. Retrieved March 15, 2024, from https://www.theyfly.com/asthetimefulfills.html

3. Horn, M. (n.d.). Blog [Review of Blog]. They Fly. Retrieved March 15, 2024, from https://www.theyfly.com/billy-meier%E2%80%99s-environmental-warnings-0 Warning to all the governments of Europe! Prophecies and Predictions (1958) item 127.

4. NASA's Moon Mission - MythBusters - S05 EP02 - Science Documentary. (n.d.). Www.youtube.com. Retrieved March 15, 2024, from https://www.youtube.com/watch?v=uGg6ywErf9Y

5. Apollo 11 Image Library. (2019). Nasa.gov. https://history.nasa.gov/alsj/a11/images11.html

6. July 16, M. R. on, & Utc, 2009 23:03. (n.d.). LROC's First Look at the Apollo Landing Sites. Www.lroc.asu.edu. Retrieved March 15, 2024, from https://www.lroc.asu.edu/posts/157

7. Villate, F. (2024, January 6). Paper [Review of Paper]. Francisco Villate Web Page. https://franciscovillate.com/docs/Apollo11-investigation-ENG.pdf

8. Horizons System. (n.d.)
   https://ssd.jpl.nasa.gov/horizons/app.html#/

9. Hasselblad. (n.d.). Www.nasa.gov. Retrieved March 15, 2024, from https://www.nasa.gov/history/alsj/a11/a11-hass.html#:~:text=And%2C%20of%20course%2C%20there%20were

10. Reseau Plate. (n.d.). Www.nasa.gov. Retrieved March 15, 2024, from https://www.nasa.gov/history/alsj/alsj-reseau.html

11. NSSDC: Apollo 11 Lunar Photography. (n.d.). Www.nasa.gov. Retrieved April 25, 2024, from https://www.nasa.gov/history/alsj/a11/a11nssdc70-06.html For Magazine 40 (S) the lens used was 60mm. For Magazine 39 (Q) the lens used was 80mm.

12. Apollo 11 Descent: Film and LRO Imagery. (n.d.). Www.youtube.com. Retrieved April 25, 2024, from https://www.youtube.com/watch?v=YKXw_3Pblh8

13. NASA. (2014). Restored Apollo 11 Moonwalk - Original NASA EVA Mission Video - Walking on the Moon [YouTube Video]. In YouTube. https://www.youtube.com/watch?v=S9HdPi9Ikhk

14. Published, R. Z. P. (2009, July 17). NASA Erased First Moonwalk Tapes, But Restores Copies. Space.com. https://www.space.com/6994-nasa-erased-moonwalk-tapes-restores-copies.html

15. NASA Releases Restored Apollo 11 Moonwalk Video - NASA. (2009, July 16). https://www.nasa.gov/missions/apollo/apollo-11/nasa-releases-restored-apollo-11-moonwalk-video/

16. Wood, B. (2005). Apollo Television [Review of Apollo Television]. https://www.nasa.gov/wp-content/uploads/static/history/alsj/ApolloTV-Acrobat7.pdf

17. From the Moon to your living room: the Apollo 11 broadcast | National Science and Media Museum. (2019, July 8). National Science and Media Museum. https://www.scienceandmediamuseum.org.uk/objects-and-stories/moon-to-living-room-apollo-11-broadcast

18. Apollo 12 - Moonwalking (Full Mission 17). (n.d.). Www.youtube.com. Retrieved March 15, 2024, from https://youtu.be/pzymTdzOAeQ?si=Dac51jgs-XAe5fBv

19. (1969, August 8). Down to the Moon [Review of Down to the Moon]. LIFE, 67(6), 18–30.

20. La foto del Apolo 11 que la NASA trucó o qué imágenes no usar para recordar a Neil Armstrong. (2012, August 26). Eureka. https://danielmarin.naukas.com/2012/08/26/la-foto-del-apolo-11-que-la-nasa-truco-o-que-imagenes-no-usar-para-recordar-a-neil-armstrong/

21. Apollo 11 Little West Crater Panorama. (n.d.). The Planetary Society. Retrieved March 15, 2024, from https://www.planetary.org/space-images/apollo-11-little-west-crater

22. Apollo 10. (n.d.). Airandspace.si.edu. https://airandspace.si.edu/explore/stories/apollo-missions/apollo-10

23. Apollo 11 Secret. (n.d.). Www.youtube.com. Retrieved April 26, 2024, from https://youtu.be/3t-OXTyGcQ8

24. Anderson, A. T. (1970). Apollo 11 Lunar Photography (C. K. Michlovitz & K. Hug, Eds.) [Review of Apollo 11 Lunar Photography]. NASA. https://ntrs.nasa.gov/api/citations/19720010768/downloads/19720010768.pdf

25. (n.d.). 70mm Hasselblad Image Catalog [Review of 70mm Hasselblad Image Catalog]. LPI; The Lunar and Planetary Institute.

26. Christopher Lock Honfsai, & Villate, F. (2020). Researching a Real UFO.

27. Apollo 13 Flight Journal - Day 1, part 2: Earth Orbit and Trans-lunar Injection. (n.d.). Www.nasa.gov. Retrieved March 15, 2024, from https://history.nasa.gov/afj/ap13fj/02earth_orbit_tli.html

28. Apollo 13 Damage Animation. (n.d.). Www.youtube.com. Retrieved March 15, 2024, from https://www.youtube.com/watch?v=44-BDXh8RqE

29. Muschalla, B., & Schönborn, F. (2021). Induction of false beliefs and false memories in laboratory studies—A systematic review. Clinical Psychology & Psychotherapy, 28(5). https://doi.org/10.1002/cpp.2567

30. Edwards, L. (2009, September). Moon Rock Turns Out to be Fake. Phys.org; Phys.org. https://phys.org/news/2009-09-moon-fake.html

31. Orloff, R. W. (2000). Apollo by the Numbers. Page 294.

LIST OF FIGURES

LIST OF TABLES

www.ingramcontent.com/pod-product-compliance
Lightning Source LLC
Chambersburg PA
CBHW040857210326
41597CB00029B/4872